ILTS Science: Physics
116 Teacher Certification Exam

By: Sharon Wynne, M.S.

XAMonline, INC.
Boston

Copyright © 2013 XAMonline, Inc.
All rights reserved. No part of the material protected by this copyright notice may be reproduced or utilized in any form or by any means, electronic or mechanical, including photocopying, recording or by any information storage and retrievable system, without written permission from the copyright holder.

To obtain permission(s) to use the material from this work for any purpose including workshops or seminars, please submit a written request to:

XAMonline, Inc.
25 First Street, Suite 106
Cambridge, MA 02141
Toll Free 1-800-301-4647
Email: info@xamonline.com
Web www.xamonline.com

Library of Congress Cataloging-in-Publication Data

Wynne, Sharon A.
　　Science: Physics 116: Teacher Certification / Sharon A. Wynne.
　　-2nd ed. ISBN 978-1-58197-999-2
　　1. Science: Physics 116.　　2. Study Guides.　　3. ILTS
　　4. Teachers' Certification & Licensure.　　5. Careers

Disclaimer:
The opinions expressed in this publication are the sole works of XAMonline and were created independently from the Pearson Corporation, National Education Association, Educational Testing Service, or any State Department of Education, National Evaluation Systems or other testing affiliates.

Between the time of publication and printing, state specific standards as well as testing formats and website information may change that is not included in part or in whole within this product. Sample test questions are developed by XAMonline and reflect similar content as on real tests; however, they are not former tests. XAMonline assembles content that aligns with state standards but makes no claims nor guarantees teacher candidates a passing score. Numerical scores are determined by testing companies such as NES or ETS and then are compared with individual state standards. A passing score varies from state to state.

Printed in the United States of America　　　　　　　　　　　　　　　　　œ-1

ILTS: Science: Physics 116
ISBN: 978-1-58197-999-2

TEACHER CERTIFICATION STUDY GUIDE

Table of Contents

SUBAREA I.   **SCIENCE AND TECHNOLOGY**

COMPETENCY 1.0 UNDERSTAND AND APPLY KNOWLEDGE OF SCIENCE AS INQUIRY 1

Skill 1.1   Recognize the assumptions, processes, purposes, requirements, and tools of scientific inquiry 1

Skill 1.2   Use evidence and logic in developing proposed explanations that address scientific questions and hypotheses 4

Skill 1.3   Identify various approaches to conducting scientific investigations and their applications 5

Skill 1.4   Use tools and mathematical and statistical methods for collecting, managing, analyzing, and communicating results of investigation 8

Skill 1.5   Demonstrate knowledge of ways to report, display, and defend the results of an investigation 9

COMPETENCY 2.0 UNDERSTAND AND APPLY KNOWLEDGE OF THE CONCEPTS, PRINCIPLES, AND PROCESSES OF TECHNOLOGICAL DESIGN 10

Skill 2.1   Recognize the capabilities, limitations, and implications of technology and technological design and redesign 10

Skill 2.2   Identify real-world problems or needs to be solved through technological design 10

Skill 2.3   Apply a technological design process to a given problem situation 10

Skill 2.4   Identify a design problem and propose possible solutions, considering such constraints as tools, materials, time, costs, and laws of nature 11

Skill 2.5   Evaluate various solutions to a design problem 11

PHYSICS

TEACHER CERTIFICATION STUDY GUIDE

COMPETENCY 3.0 UNDERSTAND AND APPLY KNOWLEDGE OF ACCEPTED PRACTICES OF SCIENCE 12

Skill 3.1 Demonstrate an understanding of the nature of science and recognize how scientific knowledge and explanations change over time.. 12

Skill 3.2 Compare scientific hypotheses, predictions, laws, theories, and principles and recognize how they are developed and tested 13

Skill 3.3 Recognize examples of valid and biased thinking in reporting of scientific research.. 15

Skill 3.4 Recognize the basis for and application of safety practices and regulations in the study of science... 15

COMPETENCY 4.0 UNDERSTAND AND APPLY KNOWLEDGE OF THE INTERACTIONS AMONG SCIENCE, TECHNOLOGY, AND SOCIETY 18

Skill 4.1 Recognize the historical and contemporary development of major scientific ideas and technological innovations 18

Skill 4.2 Demonstrate an understanding of the ways that science and technology affect people's everyday lives, societal values and systems, the environment, and new knowledge 25

Skill 4.3 Analyze the processes of scientific and technological breakthroughs and their effects on other fields of study, careers, and job markets 26

Skill 4.4 Analyze issues related to science and technology at the local, state, national, and global levels 26

Skill 4.5 Evaluate the credibility of scientific claims made in various forums 34

TEACHER CERTIFICATION STUDY GUIDE

COMPETENCY 5.0 UNDERSTAND AND APPLY KNOWLEDGE OF THE MAJOR UNIFYING CONCEPTS OF ALL SCIENCES AND HOW THESE CONCEPTS RELATE TO OTHER DISCIPLINES 35

Skill 5.1 Identify the major unifying concepts of the sciences and their applications in real-life situations 35

Skill 5.2 Recognize connections within and among the traditional scientific disciplines 36

Skill 5.3 Apply fundamental mathematical language, knowledge, and skills at the level of algebra and statistics in scientific contexts 36

Skill 5.4 Recognize the fundamental relationships among the natural sciences and the social sciences 40

PHYSICS

SUBAREA II. LIFE SCIENCE

COMPETENCY 6.0 UNDERSTAND AND APPLY KNOWLEDGE OF CELL STRUCTURE AND FUNCTION 41

Skill 6.1 Compare and contrast the structures of viruses and prokaryotic and eukaryotic cells 41

Skill 6.2 Identify the structures and functions of cellular organelles 43

Skill 6.3 Describe the processes of the cell cycle 46

Skill 6.4 Explain the functions and applications of the instruments and technologies used to study the life sciences at the molecular and cellular levels 51

COMPETENCY 7.0 UNDERSTAND AND APPLY KNOWLEDGE OF THE PRINCIPLES OF HEREDITY AND BIOLOGICAL EVOLUTION 52

Skill 7.1 Recognize the nature and function of the gene, with emphasis on the molecular basis of inheritance and gene expression 52

Skill 7.2 Analyze the transmission of genetic information 52

Skill 7.3 Analyze the processes of change at the microscopic and macroscopic levels 55

Skill 7.4 Identify scientific evidence from various sources, such as the fossil record, comparative anatomy, and biochemical similarities, to demonstrate knowledge of theories about processes of biological evolution 56

TEACHER CERTIFICATION STUDY GUIDE

COMPETENCY 8.0 UNDERSTAND AND APPLY KNOWLEDGE OF THE CHARACTERISTICS AND LIFE FUNCTIONS OF ORGANISMS 59

Skill 8.1 Identify the levels of organization of various types of organisms and the structures and functions of cells, tissues, organs, and organ systems 59

Skill 8.2 Analyze the strategies and adaptations used by organisms to obtain the basic requirements of life 59

Skill 8.3 Analyze factors that influence homeostasis within an organism 61

Skill 8.4 Demonstrate an understanding of the human as a living organism with life functions comparable to those of other life forms 61

COMPETENCY 9.0 UNDERSTAND AND APPLY KNOWLEDGE OF HOW ORGANISMS INTERACT WITH EACH OTHER AND WITH THEIR ENVIRONMENT 63

Skill 9.1 Identify living and nonliving components of the environment and how they interact with one another 63

Skill 9.2 Recognize the concepts of populations, communities, ecosystems, and ecoregions and the role of biodiversity in living systems 63

Skill 9.3 Analyze factors that influence interrelationships among organisms 63

Skill 9.4 Develop a model or explanation that shows the relationships among organisms in the environment 65

Skill 9.5 Recognize the dynamic nature of the environment, including how communities, ecosystems, and ecoregions change over time 65

Skill 9.6 Analyze interactions of humans with their environment 66

Skill 9.7 Explain the functions and applications of the instruments and technologies used to study the life sciences at the organism and ecosystem level 67

PHYSICS

TEACHER CERTIFICATION STUDY GUIDE

SUBAREA III.   **PHYSICAL SCIENCE**

COMPETENCY 10.0 UNDERSTAND AND APPLY KNOWLEDGE OF THE NATURE AND PROPERTIES OF ENERGY IN ITS VARIOUS FORMS 68

Skill 10.1   Describe the characteristics of and relationships among thermal, acoustical, radiant, electrical, chemical, mechanical, and nuclear energies through conceptual questions 68

Skill 10.2   Analyze the processes by which energy is exchanged or transformed through conceptual questions 71

Skill 10.3   Apply the three laws of thermodynamics to explain energy transformations, including basic algebraic problem solving 73

Skill 10.4   Apply the principle of conservation as it applies to energy through conceptual questions and solving basic algebraic problems 74

COMPETENCY 11.0 UNDERSTAND AND APPLY KNOWLEDGE OF THE STRUCTURE AND PROPERTIES OF MATTER 75

Skill 11.1   Describe the nuclear and atomic structure of matter, including the three basic parts of the atom 75

Skill 11.2   Analyze the properties of materials in relation to their chemical or physical structures and evaluate uses of the materials based on their properties 75

Skill 11.3   Apply the principle of conservation as it applies to mass and charge through conceptual questions 78

Skill 11.4   Analyze bonding and chemical, atomic, and nuclear reactions in natural and man-made systems and apply basic stoichiometric principles 79

Skill 11.5   Apply kinetic theory to explain interactions of energy with matter, including conceptual questions on changes in state 80

Skill 11.6   Explain the functions and applications of the instruments and technologies used to study matter and energy 80

TEACHER CERTIFICATION STUDY GUIDE

COMPETENCY 12.0 UNDERSTAND AND APPLY KNOWLEDGE OF FORCES AND MOTION 81

Skill 12.1 Demonstrate an understanding of the concepts and interrelationships of position, time, velocity, and acceleration through conceptual questions, algebra-based kinematics, and graphical analysis 81

Skill 12.2 Demonstrate an understanding of the concepts and interrelationships of force, inertia, work, power, energy, and momentum 83

Skill 12.3 Describe and predict the motions of bodies in one and two dimensions in inertial and accelerated frames of reference in a physical system, including projectile motion but excluding circular motion 85

Skill 12.4 Analyze and predict motions and interactions of bodies involving forces within the context of conservation of energy and/or momentum through conceptual questions and algebra-based problem solving 86

Skill 12.5 Describe the effects of gravitational and nuclear forces in real-life situations through conceptual questions 88

Skill 12.6 Explain the functions and applications of the instruments and technologies used to study force and motion in everyday life 89

COMPETENCY 13.0 UNDERSTAND AND APPLY KNOWLEDGE OF ELECTRICITY, MAGNETISM, AND WAVES 90

Skill 13.1 Recognize the nature and properties of electricity and magnetism, including static charge, moving charge, basic RC circuits, fields, conductors, and insulators 90

Skill 13.2 Recognize the nature and properties of mechanical an electromagnetic waves 92

Skill 13.3 Describe the effects and applications of electromagnetic forces in real-life situations, including electric power generation, circuit breakers, and brownouts 93

Skill 13.4 Analyze and predict the behavior of mechanical and electromagnetic waves under varying physical conditions, including basic optics, color, ray diagrams, and shadows 94

PHYSICS

TEACHER CERTIFICATION STUDY GUIDE

SUBAREA IV. EARTH SYSTEMS AND THE UNIVERSE

COMPETENCY 14.0 UNDERSTAND AND APPLY KNOWLEDGE OF EARTH'S LAND, WATER, AND ATMOSPHERIC SYSTEMS AND THE HISTORY OF EARTH 97

Skill 14.1 Identify the structure and composition of Earth's land, water, and atmospheric systems and how they affect weather, erosion, fresh water, and soil 97

Skill 14.2 Recognize the scope of geologic time and the continuing physical changes of Earth through time 98

Skill 14.3 Evaluate scientific theories about Earth's origin and history and how these theories explain contemporary living systems 99

Skill 14.4 Recognize the interrelationships between living organisms and Earth's resources and evaluate the uses of Earth's resources 101

COMPETENCY 15.0 UNDERSTAND AND APPLY KNOWLEDGE OF THE DYNAMIC NATURE OF EARTH 103

Skill 15.1 Analyze and explain large-scale dynamic forces, events, and processes that affect Earth's land, water, and atmospheric systems, including conceptual questions about plate tectonics, El Nino, drought, and climatic shifts 103

Skill 15.2 Identify and explain Earth processes and cycles and cite examples in real-life situations, including conceptual questions on rock cycles, volcanism, and plate tectonics 104

Skill 15.3 Analyze the transfer of energy within and among Earth's land, water, and atmospheric systems, including the identification of energy sources of volcanoes, hurricanes, thunderstorms, and tornadoes 108

Skill 15.4 Explain the functions and applications of the instruments and technologies used to study the earth sciences, including seismographs, barometers, and satellite systems 109

COMPETENCY 16.0 UNDERSTAND AND APPLY KNOWLEDGE OF OBJECTS IN THE UNIVERSE AND THEIR DYNAMIC INTERACTIONS 110

Skill 16.1 Describe and explain the relative and apparent motions of the sun, the moon, stars, and planets in the sky 110

Skill 16.2 Recognize properties of objects within the solar system and their dynamic interactions 111

Skill 16.3 Recognize the types, properties, and dynamics of objects external to the solar system 112

COMPETENCY 17.0 UNDERSTAND AND APPLY KNOWLEDGE OF THE ORIGINS OF AND CHANGES IN THE UNIVERSE 113

Skill 17.1 Identify scientific theories dealing with the origin of the universe 113

Skill 17.2 Analyze evidence relating to the origin and physical evolution of the universe 113

Skill 17.3 Compare the physical and chemical processes involved in the life cycles of objects within galaxies 114

Skill 17.4 Explain the functions and applications of the instruments, technologies, and tools used in the study of the space sciences, including the relative advantages and disadvantages of Earth-based versus space-based instruments and optical versus nonoptical instruments 116

TEACHER CERTIFICATION STUDY GUIDE

SUBAREA V.   **PHYSICS SKILLS, MOTION, FORCES, AND WAVES**

COMPETENCY 18.0 UNDERSTAND AND APPLY THE KNOWLEDGE AND SKILLS NEEDED TO PRACTICE PHYSICS AND UNDERSTAND THE BROAD APPLICABILITY OF ITS PRINCIPLES TO REAL-WORLD SITUATIONS 117

Skill 18.1   Demonstrate knowledge of the safe and proper use of equipment and materials commonly used in physics classrooms and laboratories 117

Skill 18.2   Design appropriate laboratory investigations to study the principles and applications of physics 119

Skill 18.3   Demonstrate knowledge of the uses of basic equipment to illustrate physical principles and phenomena 120

Skill 18.4   Use mathematical concepts, strategies, and procedures, including graphical and statistical methods and differential and integral calculus, to derive and manipulate formal relationships between physical quantities 122

Skill 18.5   Demonstrate an understanding of the growth of physics knowledge from a historical perspective 125

COMPETENCY 19.0 UNDERSTAND AND APPLY KNOWLEDGE OF PLANAR MOTION ... 129

Skill 19.1 Analyze the relationship between vectors and physical quantities and perform a variety of vector algebra operations ... 129

Skill 19.2 Use algebra and calculus methods to determine the rectilinear displacement, velocity, and acceleration of particles and rigid bodies, given initial conditions ... 132

Skill 19.3 Use algebra and calculus methods to determine angular displacement, velocity, and acceleration of rigid bodies in a plane, given initial conditions ... 134

Skill 19.4 Use algebra and calculus methods to determine the displacement, velocity, and acceleration of particles and rigid bodies undergoing periodic motion, given initial conditions ... 137

Skill 19.5 Analyze and solve problems involving the relationships of linear and angular displacement, velocity and acceleration ... 138

Skill 19.6 Analyze and solve problems involving periodic motion and uniform circular motion ... 141

COMPETENCY 20.0 UNDERSTAND AND APPLY KNOWLEDGE OF FORCE, MOMENTUM, AND ENERGY AS THEY APPLY TO PLANAR MOTION ... 142

Skill 20.1 Apply Newton's laws of motion to analyze and solve problems involving translational, rotational, and periodic motion ... 142

Skill 20.2 Apply the law of universal gravitation to solve problems involving free fall, projectile motion, and planetary motion ... 143

Skill 20.3 Analyze and solve problems involving the relationships between linear quantities and their rotational analogues ... 144

Skill 20.4 Solve problems involving the conservation of linear and angular momentum ... 146

Skill 20.5 Use the relationship between work and energy, in algebraic and calculus forms, to solve problems involving the motions of physical systems acted upon by conservative and nonconservative forces ... 147

COMPETENCY 21.0	UNDERSTAND AND APPLY KNOWLEDGE OF THE NATURE, PROPERTIES, AND BEHAVIOR OF MECHANICAL WAVES 149
Skill 21.1	Apply the relationships among wave speed, wavelength, period, and frequency to analyze and solve problems related to wave propagation 149
Skill 21.2	Analyze the interference and reflection of waves and wave pulses 151
Skill 21.3	Describe and analyze the nature, production and transmission of sound waves in various uniform media 152
Skill 21.4	Describe how the perception of sound depends on the physical properties of sound waves 154

COMPETENCY 22.0	UNDERSTAND AND APPLY KNOWLEDGE OF THE NATURE, PROPERTIES, AND BEHAVIOR OF ELECTROMAGNETIC RADIATION 156
Skill 22.1	Classify the regions of the electromagnetic spectrum relative to their frequency or wavelength 156
Skill 22.2	Analyze and predict the behavior of various types of electromagnetic radiation as they interact with matter 156
Skill 22.3	Analyze and predict the behaviors of light, including interference, reflection, diffraction, polarization, and refraction 159
Skill 22.4	Use ray diagrams to analyze systems of lenses and mirrors 164

TEACHER CERTIFICATION STUDY GUIDE

SUBAREA VI. HEAT, ELECTRICITY, MAGNETISM, AND MODERN PHYSICS

COMPETENCY 23.0   **UNDERSTAND AND APPLY KNOWLEDGE OF THE PRINCIPLES OF THERMODYNAMICS** 167

Skill 23.1   Apply basic concepts of heat and temperature as they relate to temperature measurement and temperature-dependent properties of matter 167

Skill 23.2   Apply the laws of thermodynamics to problems involving temperature, work, heat, energy, and entropy 168

Skill 23.3   Demonstrate knowledge of the kinetic-molecular theory and apply it to describe thermal properties and behaviors of solids, liquids and gases 171

Skill 23.4   Analyze and solve problems involving energy, temperature, heat, and changes of state 172

COMPETENCY 24.0   **UNDERSTAND AND APPLY KNOWLEDGE OF STATIC AND MOVING ELECTRIC CHARGES** 177

Skill 24.1   Predict the interactions between electric charges 177

Skill 24.2   Interpret electric field diagrams and predict the influence of electric fields on electric charges 179

Skill 24.3   Determine the electric potential due to a charge distribution and calculate the work involved in moving a point charge through a potential difference 180

Skill 24.4   Determine the electric field due to a charge distribution and calculate the force on a point charge located in that electric field 183

Skill 24.5   Describe the flow of charge through different media and interpret circuit diagrams 185

Skill 24.6   Analyze AC and DC circuits composed of basic circuit elements 187

PHYSICS

COMPETENCY 25.0 UNDERSTAND AND APPLY KNOWLEDGE OF THE PRINCIPLES OF MAGNETISM AND INDUCED ELECTRIC FIELDS 191

Skill 25.1 Analyze the motion of a charged particle in a magnetic field and determine the force on a current carrying conductor in a magnetic field 191

Skill 25.2 Analyze the characteristics of magnetic fields produced by straight and coiled current-carrying conductors 195

Skill 25.3 Describe and analyze the processes of electromagnetic induction 197

Skill 25.4 Demonstrate an understanding of the operating principles of electric generators, motors and transformers 198

Skill 25.5 Identify applications of magnets and magnetic fields in technology and daily living 200

TEACHER CERTIFICATION STUDY GUIDE

COMPETENCY 26.0   **UNDERSTAND AND APPLY KNOWLEDGE OF THE BASIC CONCEPTS AND APPLICATIONS OF MODERN PHYSICS** **202**

Skill 26.1   Demonstrate knowledge of a quantum model of atomic structure, including the relationship between changes in electron energy levels and atomic spectra 202

Skill 26.2   Describe types, properties, and applications of radioactivity and the effects of radioactivity on living organisms 204

Skill 26.3   Balance particle equations and solve radioactive decay problems involving half-life, energy, mass and charge 205

Skill 26.4   Describe the quantum mechanical nature of the interaction between radiation and matter 208

Skill 26.5   Describe the wave-particle duality of radiation and matter 209

Skill 26.6   Describe the quantum mechanical electron properties of conductors, semiconductors, and insulators. 210

Skill 26.7   Apply the concepts of special relativity as they relate to time, space and mass 212

Sample Test **215**

Answer Key **230**

Rationales with Sample Questions **231**

PHYSICS   xv

TEACHER CERTIFICATION STUDY GUIDE

Great Study and Testing Tips!

What to study in order to prepare for the subject assessments is the focus of this study guide but equally important is *how* you study.

You can increase your chances of truly mastering the information by taking some simple, but effective steps.

Study Tips:

1. Some foods aid the learning process. Foods such as milk, nuts, seeds, rice, and oats help your study efforts by releasing natural memory enhancers called CCKs (*cholecystokinin*) composed of *tryptophan*, *choline*, and *phenylalanine*. All of these chemicals enhance the neurotransmitters associated with memory. Before studying, try a light, protein-rich meal of eggs, turkey, and fish. All of these foods release the memory enhancing chemicals. The better the connections, the more you comprehend.

Likewise, before you take a test, stick to a light snack of energy boosting and relaxing foods. A glass of milk, a piece of fruit, or some peanuts all release various memory-boosting chemicals and help you to relax and focus on the subject at hand.

2. Learn to take great notes. A by-product of our modern culture is that we have grown accustomed to getting our information in short doses (i.e. TV news sound bites or USA Today style newspaper articles.)

Consequently, we've subconsciously trained ourselves to assimilate information better in neat little packages. If your notes are scrawled all over the paper, it fragments the flow of the information. Strive for clarity. Newspapers use a standard format to achieve clarity. Your notes can be much clearer through use of proper formatting. A very effective format is called the "Cornell Method."

Take a sheet of loose-leaf lined notebook paper and draw a line all the way down the paper about 1-2" from the left-hand edge.

Draw another line across the width of the paper about 1-2" up from the bottom. Repeat this process on the reverse side of the page.

Look at the highly effective result. You have ample room for notes, a left hand margin for special emphasis items or inserting supplementary data from the textbook, a large area at the bottom for a brief summary, and a little rectangular space for just about anything you want.

PHYSICS

3. Get the concept then the details. Too often we focus on the details and don't gather an understanding of the concept. However, if you simply memorize only dates, places, or names, you may well miss the whole point of the subject. A key way to understand things is to put them in your own words. If you are working from a textbook, automatically summarize each paragraph in your mind. If you are outlining text, don't simply copy the author's words.

Rephrase them in your own words. You remember your own thoughts and words much better than someone else's, and subconsciously tend to associate the important details to the core concepts.

4. Ask Why? Pull apart written material paragraph by paragraph and don't forget the captions under the illustrations.

Example: If the heading is "Stream Erosion", flip it around to read "Why do streams erode?" Then answer the questions.

If you train your mind to think in a series of questions and answers, not only will you learn more, but it also helps to lessen the test anxiety because you are used to answering questions.

5. Read for reinforcement and future needs. Even if you only have 10 minutes, put your notes or a book in your hand. Your mind is similar to a computer; you have to input data in order to have it processed. *By reading, you are creating the neural connections for future retrieval.* The more times you read something, the more you reinforce the learning of ideas.

Even if you don't fully understand something on the first pass, *your mind stores much of the material for later recall.*

6. Relax to learn so go into exile. Our bodies respond to an inner clock called biorhythms. Burning the midnight oil works well for some people, but not everyone.

If possible, set aside a particular place to study that is free of distractions. Shut off the television, cell phone, pager and exile your friends and family during your study period.

If you really are bothered by silence, try background music. Light classical music at a low volume has been shown to aid in concentration over other types.

Music that evokes pleasant emotions without lyrics are highly suggested. Try just about anything by Mozart. It relaxes you.

7. Use arrows not highlighters. At best, it's difficult to read a page full of yellow, pink, blue, and green streaks.

Try staring at a neon sign for a while and you'll soon see my point, the horde of colors obscure the message.

8. Budget your study time. Although you shouldn't ignore any of the material, *allocate your available study time in the same ratio that topics may appear on the test.*

TEACHER CERTIFICATION STUDY GUIDE

Testing Tips:

1. Get smart, play dumb. Don't read anything into the question. Don't make an assumption that the test writer is looking for something else than what is asked. Stick to the question as written and don't read extra things into it.

2. Read the question and all the choices *twice* before answering the question. You may miss something by not carefully reading, and then re-reading both the question and the answers.

If you really don't have a clue as to the right answer, leave it blank on the first time through. Go on to the other questions, as they may provide a clue as to how to answer the skipped questions.

If later on, you still can't answer the skipped ones . . . ***Guess.*** The only penalty for guessing is that you *might* get it wrong. Only one thing is certain; if you don't put anything down, you will get it wrong!

3. Turn the question into a statement. Look at the way the questions are worded. The syntax of the question usually provides a clue. Does it seem more familiar as a statement rather than as a question? Does it sound strange?

By turning a question into a statement, you may be able to spot if an answer sounds right, and it may also trigger memories of material you have read.

4. Look for hidden clues. It's actually very difficult to compose multiple-foil (choice) questions without giving away part of the answer in the options presented.

In most multiple-choice questions you can often readily eliminate one or two of the potential answers. This leaves you with only two real possibilities and automatically your odds go to Fifty-Fifty for very little work.

5. Trust your instincts. For every fact that you have read, you subconsciously retain something of that knowledge. On questions that you aren't really certain about, go with your basic instincts. **Your first impression on how to answer a question is usually correct.**

6. Mark your answers directly on the test booklet. Don't bother trying to fill in the optical scan sheet on the first pass through the test.

Just be very careful not to miss-mark your answers when you eventually transcribe them to the scan sheet.

7. Watch the clock! You have a set amount of time to answer the questions. Don't get bogged down trying to answer a single question at the expense of 10 questions you can more readily answer.

PHYSICS

TEACHER CERTIFICATION STUDY GUIDE

SUBAREA I. SCIENCE AND TECHNOLOGY

COMPETENCY 1.0 UNDERSTAND AND APPLY KNOWLEDGE OF SCIENCE AS INQUIRY

Skill 1.1 Recognize the assumptions, processes, purposes, requirements, and tools of scientific inquiry.

Modern science began around the late 16th century with a new way of thinking about the world. Few scientists will disagree with Carl Sagan's assertion that "science is a way of thinking much more than it is a body of knowledge" (Broca's Brain, 1979). Thus science is a process of inquiry and investigation. It is a way of thinking and acting, not just a body of knowledge to be acquired by memorizing facts and principles. This way of thinking, the scientific method, is based on the idea that scientists begin their investigations with observations. From these observations they develop a hypothesis, which is extended in the form of a predication, and challenge the hypothesis through experimentation and thus further observations. Science has progressed in its understanding of nature through careful observation, a lively imagination, and increasing sophisticated instrumentation. Science is distinguished from other fields of study in that it provides guidelines or methods for conducting research, and the research findings must be reproducible by other scientists for those findings to be valid. It is important to recognize that scientific practice is not always this systematic. Discoveries have been made that are serendipitous and others have not started with the observation of data. Einstein's theory of relativity started not with the observation of data but with a kind of intellectual puzzle.

The Scientific method is just a logical set of steps that a scientist goes through to solve a problem. There are as many different scientific methods as there are scientists experimenting. However, there seems to be some pattern to their work.

While an inquiry may start at any point in this method and may not involve all of the steps here is the pattern.

Observations
Scientific questions result from observations of events in nature or events observed in the laboratory. An **observation** is not just a look at what happens. It also includes measurements and careful records of the event. Records could include photos, drawings, or written descriptions. The observations and data collection lead to a question. In chemistry, observations almost always deal with the behavior of matter. Having arrived at a question, a scientist usually researches the scientific literature to see what is known about the question. Maybe the question has already been answered. The scientist then may want to test the answer found in the literature. Or, maybe the research will lead to a new question.

PHYSICS

Sometimes the same observations are made over and over again and are always the same. For example, you can observe that daylight lasts longer in summer than in winter. This observation never varies. Such observations are called **laws** of nature. Probably the most important law in chemistry was discovered in the late 1700s. Chemists observed that no mass was ever lost or gained in chemical reactions. This law became known as the law of conservation of mass. Explaining this law was a major topic of chemistry in the early 19th century.

Hypothesis

If the question has not been answered, the scientist may prepare for an experiment by making a hypothesis. A **hypothesis** is a statement of a possible answer to the question. It is a tentative explanation for a set of facts and can be tested by experiments. Although hypotheses are usually based on observations, they may also be based on a sudden idea or intuition.

Experiment

An **experiment** tests the hypothesis to determine whether it may be a correct answer to the question or a solution to the problem. Some experiments may test the effect of one thing on another under controlled conditions. Such experiments have two variables. The experimenter controls one variable, called the *independent variable*. The other variable, the *dependent variable*, is the change caused by changing the independent variable.

For example, suppose a researcher wanted to test the effect of vitamin A on the ability of rats to see in dim light. The independent variable would be the dose of Vitamin A added to the rats' diet. The dependent variable would be the intensity of light that causes the rats to react. All other factors, such as time, temperature, age, water given to the rats, the other nutrients given to the rats, and similar factors, are held constant. Chemists sometimes do short experiments "just to see what happens" or to see what products a certain reaction produces. Often, these are not formal experiments. Rather they are ways of making additional observations about the behavior of matter.

In most experiments, scientists collect quantitative data, which is data that can be measured with instruments. They also collect qualitative data, descriptive information from observations other than measurements. Interpreting data and analyzing observations are important. If data is not organized in a logical manner, wrong conclusions can be drawn. Also, other scientists may not be able to follow your work or repeat your results.

Steps of a Scientific Method

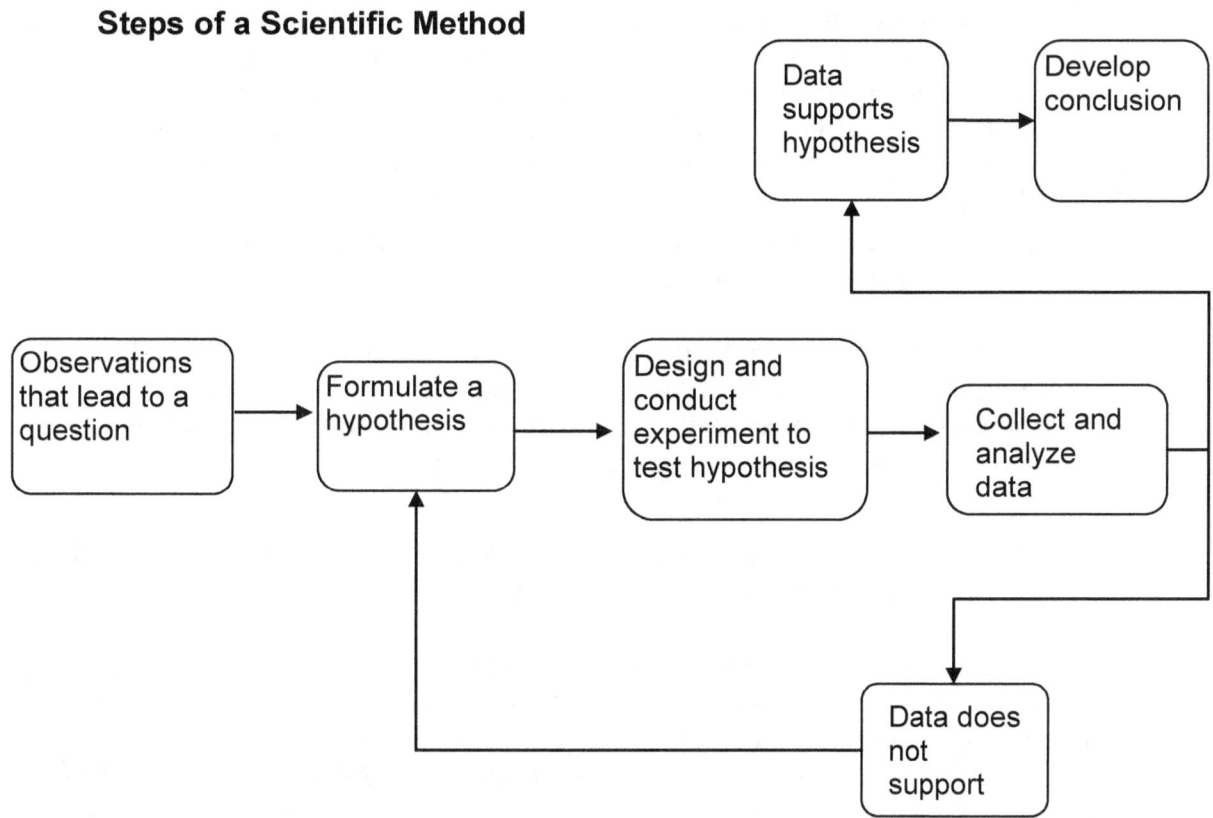

Conclusion

Finally, a scientist must draw conclusions from the experiment. A conclusion must address the hypothesis on which the experiment was based. The conclusion states whether or not the data supports the hypothesis. If it does not, the conclusion should state what the experiment *did* show. If the hypothesis is not supported, the scientist uses the observations from the experiment to make a new or revised hypothesis. Then, new experiments are planned.

Theory

When a hypothesis survives many experimental tests to determine its validity, the hypothesis may evolve into a **theory**. A theory explains a body of facts and laws that are based on the facts. A theory also reliably predicts the outcome of related events in nature. For example, the law of conservation of matter and many other experimental observations led to a theory proposed early in the 19th century. This theory explained the conservation law by proposing that all matter is made up of atoms which are never created or destroyed in chemical reactions, only rearranged. This atomic theory also successfully predicted the behavior of matter in chemical reactions that had not been studied at the time. As a result, the atomic theory has stood for 200 years with only small modifications.

A theory also serves as a scientific **model**. A model can be a physical model made of wood or plastic, a computer program that simulates events in nature, or simply a mental picture of an idea. A model illustrates a theory and explains nature. In your chemistry course, you will develop a mental (and maybe a physical) model of the atom and its behavior. Outside of science, the word theory is often used to describe someone's unproven notion about something. In science, theory means much more. It is a thoroughly tested explanation of things and events observed in nature.

A theory can never be proven true, but it can be proven untrue. All it takes to prove a theory untrue is to show an exception to the theory. The test of the hypothesis may be observations of phenomena or a model may be built to examine its behavior under certain circumstances.

Skill 1.2   Use evidence and logic in developing proposed explanations that address scientific questions and hypotheses.

A scientific model is a set of ideas that describes a natural process and are developed by empirical or theoretical objects and help scientists to focus on the basic fundamental processes. They may be physical representations such as a space-filling model of a molecule or a map, or they may be mathematical algorithms.

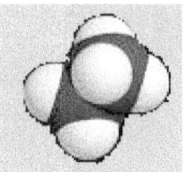

Whatever form they take, scientific models are based on what is known about the science systems or objects at the time that the models are constructed. Models usually evolve and are improved as scientific advances are made. Sometimes, a model must be discarded because new findings show it to be misleading or incorrect. How do scientists use models?

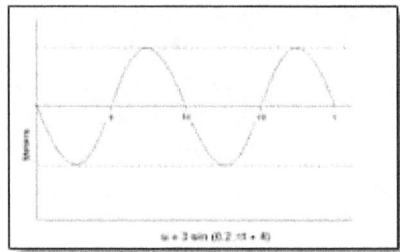

Models are developed in an effort to explain how things work in nature. Because models are not the "real thing", they can never correctly represent the system or object in all respects. The amount of detail that they contain depends upon how the model will be used as well as the sophistication and skill of the scientist doing the modeling. If a model has too many details left out, its usefulness may be limited. But too many details may make a model too complicated to be useful. So it is easy to see why models lack some features of the real system.

To overcome this difficulty, different models are often used to describe the same system or object. Scientists must then choose which model most closely fits the scientific investigation being carried out, which findings are being described, and, in some cases, which one is compatible with the sophistication of the investigation itself. For example, there are many models of atoms. The solar system model described above is adequate for some purposes, because electrons have properties of matter. They have mass and charge and they are found in motion in the space outside the nucleus. However, a highly mathematical model based on the field of quantum mechanics is necessary when describing the energy (or wave) properties of electrons in the atom.

Scientific models are based on physical observations that establish some facts about the system or object of interest. Scientists then combine these facts with appropriate laws or scientific principles and assumptions to produce a "picture" that mimics the behavior of the system or object to the greatest possible extent. It is on the basis of such models that science makes many of its most important advances, because such models provide a vehicle for making predictions about the behavior of a system or object. The predictions can then be tested as new measurements, technology or theories are applied to the subject. The new information may result in modification and refinement of the model, although certain issues may remain unresolved by the model for years. The goal, however is to continue to develop the model in such a way as to move it every closer to a true description of the natural phenomenon. In this way, models are vital to the scientific process.

Skill 1.3 Identify various approaches to conducting scientific investigations and their applications.

The design of chemical experiments must include every step to obtain the desired data. In other words, the design must be **complete** and it must include all required **controls**.

Complete design
Familiarity with individual experiments and equipment will help you evaluate if anything is missing from the design. For data requiring a difference between two values, the experiment **must determine both values**. For data utilizing the ideal gas law, the experiment **must determine three values of P, V, n, or T** in order to determine the fourth or one value and a ratio of the other two in order to determine the fourth.

Example: In a mercury manometer, the level of mercury in contact with a reaction vessel is 70.0 mm lower than the level exposed to the atmosphere. Use the following conversion factors:
$$760 \text{ mm Hg} = 1 \text{ atm} = 101.325 \text{ kPa}.$$
What additional information is required to determine the pressure in the vessel in kPa?

Solution: The barometric pressure is needed to determine vessel pressure from an open-ended manometer. A manometer reading is always a **difference** between two pressures. One standard atmosphere is 760 mm mercury, but on a given day at a given location, the actual ambient pressure may vary. If the barometric pressure on the day of the experiment is 104 kPa, the pressure of the vessel is:

$$104 \text{ kPa} + 70.0 \text{ mm Hg} \times \frac{101.325 \text{ kPa}}{760 \text{ mm Hg}} = 113 \text{ kPa}.$$

Controls: Experimental **controls** prevent factors other than those under study from impacting the outcome of the experiment. An **experimental sample** in a controlled experiment is the unknown to be compared against one or more **control samples**. These should be nearly identical to the experimental sample except for the one aspect whose effect is being tested.

A **negative control** is a control sample that is known to lack the effect. A **positive control** is known to contain the effect. Positive controls of varying strengths are often used to generate a **calibration curve** (also called a **standard curve**).

When determining the concentration of a component in a mixture, an **internal standard** is a known concentration of a different substance that is added to the experimental sample. An **external standard** is a known concentration of the substance of interest. External standards are more commonly used. They are not added to the experimental sample; they are analyzed separately.

Replicate samples decrease the impact of random error. A mean is taken of the results from replicate samples to obtain a best value. If one replicate is obviously inconsistent with the results from other samples, it may be discarded as an **outlier** and not counted as an observation when determining the mean. Discarding an outlier is equivalent to assuming the presence of a systematic error for that particular observation. In research, this must be done with great caution because some real-world behavior generates sporadically unusual results.

Example: A pure chemical in aqueous solution is known to absorb light at 615 nm. What controls would best be used with a spectrophotometer to determine the concentration of this chemical when it is present in a mixture with other solutes in an aqueous solution?

Solution: The other solutes may also absorb light at 615 nm. The best negative control would be an identical mixture with the chemical of interest entirely absent. Known concentrations of the chemical could then be added to the negative control to create positive controls (external standards) and develop a calibration curve of the spectrophotometer absorbance reading at 615 nm as a function of concentration. Replicate samples of each standard and of the unknown should be read.

Example: Ethanol is separated from a mixture of organic compounds by gas chromatography. The concentration of each component is proportional to its peak area. However, the chromatograph detector has a variable sensitivity from one run to the next. Is an internal standard required to determine the concentration of ethanol?

Solution: Yes. The variable detector sensitivity may only be accounted for by adding a known concentration of a chemical not found in the mixture as an internal standard to the experimental sample and control samples. The variable sensitivity of the detector will be accounted for by determining the ratio of the peak area for ethanol to the peak area of the added internal standard.

Experimental bias is when a researcher favors one particular outcome over another in an experimental setup. In order to avoid bias, it is imperative to set each experiment under exactly the same conditions, including a *control* experiment, an experiment with a known negative outcome. Additionally, in order to avoid experimental bias, a researcher must not "read" particular results into data.

An example of experimental bias can be seen in the classic example of the mouse in the maze experiment. In this example, a researcher is timing mice as they move through the maze towards a piece of cheese. The experiment relies on the mouse's ability to smell the cheese as it approaches. If one mouse chases a piece of cheddar cheese, while another chases Limburger, or so called "stinky" cheese, clearly the Limburger mouse has a huge advantage over the cheddar mouse. To remove the experimental bias from this experiment, the same cheese should be used in both tests.

Skill 1.4 Use tools and mathematical and statistical methods for collecting, managing, analyzing (e.g., average, curve fit, error determination), and communicating results of investigations.

The knowledge and use of basic mathematical concepts and skills is a necessary aspect of scientific study. Science depends on data and the manipulation of data requires knowledge of mathematics. Scientists often use basic algebra to solve scientific problems and design experiments. For example, the substitution of variables is a common strategy in experiment design. Also, the ability to determine the equation of a curve is valuable in data manipulation, experimentation, and prediction.

Understanding of basic statistics, graphs and charts, and algebra are of particular importance. In addition, scientists must be able to represent data graphically and interpret graphs and tables.

Scientists must be able to understand and apply the statistical concepts of mean, median, mode, and range to sets of scientific data. Modern science uses a number of disciplines to understand it better. Statistics is one of those subjects, which is absolutely essential for science.

Mean is the mathematical average of all the items. To calculate the mean, all the items must be added up and divided by the number of items. This is also called the arithmetic mean or more commonly as the "**average**".

The **median** depends on whether the number of items is odd or even. If the number is odd, then the median is the value of the item in the middle. This is the value that denotes that the number of items having higher or equal value to that is same as the number of items having equal or lesser value than that. If the number of the items is even, the median is the average of the two items in the middle, such that the number of items having values higher or equal to it is same as the number of items having values equal or less than that.

Mode is the value of the item that occurs the most often, if there are not many items. Bimodal is a situation where there are two items with equal frequency.

Range is the difference between the maximum and minimum values. The range is the difference between two extreme points on the **distribution curve**.

Error

There are many ways in which errors could creep in measurements.
Errors in measurements could occur because –

1. Improper use of instruments used for measuring – weighing etc.
2. Parallax error – not positioning the eyes during reading of measurements
3. Not using same instruments and methods of measurement during an experiment
4. Not using the same source of materials, resulting in the content of a certain compound used for experimentation

Besides these mentioned above, there could be other possible sources of error as well. When erroneous results are used for interpreting data, the conclusions are not reliable. An experiment is valid only when all the constants like time, place, method of measurement, etc. are strictly controlled.

Skill 1.5   Demonstrate knowledge of ways to report, display, and defend the results of an investigation.

Lab notebooks are commonly used in the laboratory to record data. This information is often then transferred into a computer file or spreadsheet once outside of the lab. Scientists use spreadsheets to organize, analyze, and display data. Use of spreadsheets simplifies data collection and manipulation and allows the presentation of data in a logical and understandable format. Models are another common way of portraying evidence. A model may be built to demonstrate an object too large or too small to otherwise visualize with ease. In the case of chemistry, computers and their simulations are often used to show the three dimensional structure of a molecule and to predict its possible chemical interactions.

It is the responsibility of the scientists to share the knowledge they obtain through their research. After the conclusion is drawn, the final step is communication. In this age, much emphasis is put on the way and the method of communication. The conclusions must be communicated by clearly describing the information using accurate data, visual presentation and other appropriate media such as a power point presentation. Examples of visual presentations are graphs (bar/line/pie), tables/charts, diagrams, and artwork. Modern technology must be used whenever necessary. The method of communication must be suitable to the audience.

Written communication is as important as oral communication. This is essential for submitting research papers to scientific journals, newspapers, other magazines etc.

COMPETENCY 2.0 UNDERSTAND AND APPLY KNOWLEDGE OF THE CONCEPTS, PRINCIPLES, AND PROCESSES OF TECHNOLOGICAL DESIGN.

Skill 2.1 Recognize the capabilities, limitations, and implications of technology and technological design and redesign.

Science and technology are interdependent as advances in technology often lead to new scientific discoveries and new scientific discoveries often lead to new technologies. Scientists use technology to enhance the study of nature and solve problems that nature presents. Technological design is the identification of a problem and the application of scientific knowledge to solve the problem.

While technology and technological design can provide solutions to problems faced by humans, technology must exist within nature and cannot contradict physical or biological principles. In addition, technological solutions are temporary and new technologies typically provide better solutions in the future. Monetary costs, available materials, time, and available tools also limit the scope of technological design and solutions. Finally, technological solutions have intended benefits and unexpected consequences. Scientists must attempt to predict the unintended consequences and minimize any negative impact on nature or society.

Skill 2.2 Identify real-world problems or needs to be solved through technological design.

The problems and needs, ranging from very simple to highly complex, that technological design can solve are nearly limitless. Disposal of toxic waste, routing of rainwater, crop irrigation, and energy creation are but a few examples of real-world problems that scientists address or attempt to address with technology.

Skill 2.3 Apply a technological design process to a given problem situation.

The technological design process has five basic steps:
1. Identify a problem
2. Propose designs and choose between alternative solutions
3. Implement the proposed solution
4. Evaluate the solution and its consequences
5. Report results

After the identification of a problem, the scientist must propose several designs and choose between the alternatives. Scientists often utilize simulations and models in evaluating possible solutions.

Implementation of the chosen solution involves the use of various tools depending on the problem, solution, and technology. Scientists may use both physical tools and objects and computer software.

After implementation of the solution, scientists evaluate the success or failure of the solution against pre-determined criteria. In evaluating the solution, scientists must consider the negative consequences as well as the planned benefits.

Finally, scientists must communicate results in different ways – orally, written, models, diagrams, and demonstrations.

Example:

Problem – toxic waste disposal
Chosen solution – genetically engineered microorganisms to digest waste
Implementation – use genetic engineering technology to create organism capable of converting waste to environmentally safe product
Evaluate – introduce organisms to waste site and measure formation of products and decrease in waste; also evaluate any unintended effects
Report – prepare a written report of results complete with diagrams and figures.

Skill 2.4 Identify a design problem and propose possible solutions, considering such constraints as tools, materials, time, costs, and laws of nature.

In addition to finding viable solutions to design problems, scientists must consider such constraints as tools, materials, time, costs, and laws of nature. Effective implementation of a solution requires adequate tools and materials. Scientists cannot apply scientific knowledge without sufficient technology and appropriate materials (e.g. construction materials, software). Technological design solutions always have costs. Scientists must consider monetary costs, time costs, and the unintended effects of possible solutions. Types of unintended consequences of technological design solutions include adverse environmental impact and safety risks. Finally, technology cannot contradict the laws of nature. Technological design solutions must work within the framework of the natural world.

Skill 2.5 Evaluate various solutions to a design problem.

In evaluating and choosing between potential solutions to a design problem, scientists utilize modeling, simulation, and experimentation techniques. Small-scale modeling and simulation help test the effectiveness and unexpected consequences of proposed solutions while limiting the initial costs. Modeling and simulation may also reveal potential problems that scientists can address prior to full-scale implementation of the solution. Experimentation allows for evaluation of proposed solutions in a controlled environment where scientists can manipulate and test specific variables.

COMPETENCY 3.0 UNDERSTAND AND APPLY KNOWLEDGE OF ACCEPTED PRACTICES OF SCIENCE.

Skill 3.1 Demonstrate an understanding of the nature of science (e.g., tentative, replicable, historical, empirical) and recognize how scientific knowledge and explanations change over time.

Probably one of the best examples of the progressive development of science would be the development of atomic theory. The ancient Greeks debated over the continuous nature of matter and two schools of thought emerged; matter is continuous or matter is not continuous. The continuous idea was promoted by Aristotle and due to his high regard among scholars that was the idea that flourished. However, there was no effort made by the Greeks to prove of disprove this idea. During the Dark Ages alchemists started experimenting and keeping records of their results, sending science on a pathway of experimentation and discovery. Robert Boyle and his famous J-tube experiment in 1661 gave the first experimental evidence for the existence of atoms. He even used words similar to Democritus to describe the results saying that the air consisted of atoms and a void between them. By increasing the pressure inside the J-tube, some of the void was squeezed out, decreasing the volume. Slowly, experimental evidence, including the work of Lavoisier and Priestly, to name a few, began to mount and in 1803 John Dalton proposed the Modern Atomic Theory which contained 5 basic postulates about the nature and behavior of matter. Ben Franklin's discovery of electricity in 1746 sent scientists to work to understand this new "thing"-electricity. J.J Thomson investigated a cathode ray tube and identified the negatively charged particle in the cathode ray in 1897. His work was closely followed by Robert Milliken who gave the electron a -1 charge in his oil drop experiment.

Experiments were under way to understand how electricity and matter interact when the discoveries of x-rays and radioactivity were announced. Scientists trying to understand this new phenomena radioactivity experimented day and night. Ernest Rutherford was one of the many. He tried to understand the nature of radioactivity and classified it into three basic types. While trying to find out more about radioactivity, he conducted his gold foil experiment that ultimately provided greater insight into the subatomic nature of the atom by discovering the nucleus. He also identified the proton present in the nucleus.

Rutherford's graduate student, Neils Bohr, made slight alterations to the model of the atom proposed by Rutherford to account for his experimental results. These changes helped Rutherford's model stand up to classical physics. However, scientists looking for other patterns and information proposed more changes to the planetary model of the atom. These changes seemed to fit with spectroscopy experiments and the quantum mechanical view of the atom was formed. About the same time as this new theory was emerging, James Chadwick, a collaborator of Rutherford's, announced in 1932 the discovery of the neutron found in the nucleus. This discovery, ultimately led to the discovery of fission and the development of the nuclear bomb. The technology the came from the Manhattan project has developed the fields of medicine, computing and provided entertainment from television.

Skill 3.2 Compare scientific hypotheses, predictions, laws, theories, and principles and recognize how they are developed and tested.

A Law is the highest-level science can achieve. Followed by laws and theories and hypothesis. The Scientific Method is the process by which data is collected, interpreted and validated.

Law is defined as: a statement of an order or relation of phenomena that so far as is known is invariable under the given conditions. Everything we observe in the universe operates according to known natural laws.

- If the truth of a statement is verified repeatedly in a reproducible way then it can reach the level of a natural law.

- Some well know and accepted natural laws of science are:

    1. The First Law of Thermodynamics

    2. The Second Law of Thermodynamics

    3. The Law of Cause and Effect

    4. The Law of Biogenesis

    5. The Law of Gravity

Theory is defined as: In contrast to a law, a scientific theory is used to explain an observation or a set of observations. It is generally accepted to be true, though no real proof exists. The important thing about a scientific law is that there are no experimental observations to prove it NOT true, and each piece of evidence that exists supports the theory as written. They are often accepted at face value, since they are often difficult to prove, and can be rewritten in order include the results of all experimental observations. An example of a theory is the big bang theory. While there is no experiment that can directly test whether or not the big bang actually occurred, there is no strong evidence indicating otherwise.

Theories provide a framework to explain the **known** information of the time, but are subject to constant evaluation and updating. There is always the possibility that new evidence will conflict with a current theory.

Some examples of theories that have been rejected because they are now better explained by current knowledge:

Theory of Spontaneous Generation
Inheritance of Acquired Characteristics
The Blending Hypothesis

Some examples of theories that were initially rejected because they fell outside of the accepted knowledge of the time, but are well-accepted today due to increased knowledge and data include:

The sun-centered solar system
Warm-bloodedness in dinosaurs
The germ-theory of disease
Continental drift

Hypothesis is defined as: a tentative assumption made in order to draw out and test its logical or empirical consequences. Many refer to an hypothesis as an educated guess about what will happen during an experiment. A hypothesis can be based on prior knowledge, prior observations. It will be proved true or false only through experimentation.

Scientific Method is defined as: principles and procedures for the systematic pursuit of knowledge involving the recognition and formulation of a problem, the collection of data through observation and experiment, and the formulation and testing of hypotheses. The steps in the scientific method can be found elsewhere in this text.

Skill 3.3 **Recognize examples of valid and biased thinking in reporting of scientific research.**

Scientific research can be biased in the choice of what data to consider, in the reporting or recording of the data, and/or in how the data are interpreted. The scientist's emphasis may be influenced by his/her nationality, sex, ethnic origin, age, or political convictions. For example, when studying a group of animals, male scientists may focus on the social behavior of the males and typically male characteristics. Although bias related to the investigator, the sample, the method, or the instrument may not be completely avoidable in every case, it is important to know the possible sources of bias and how bias could affect the evidence. Moreover, scientists need to be attentive to possible bias in their own work as well as that of other scientists.

Objectivity may not always be attained. However, one precaution that may be taken to guard against undetected bias is to have many different investigators or groups of investigators working on a project. By different, it is meant that the groups are made up of various nationalities, ethnic origins, ages, and political convictions and composed of both males and females. It is also important to note one's aspirations, and to make sure to be truthful to the data, even when grants, promotions, and notoriety are at risk.

Skill 3.4 **Recognize the basis for and application of safety practices and regulations in the study of science.**

All science labs should contain the following items of **safety equipment**. Those marked with an asterisk are requirements by state laws.

* fire blanket which is visible and accessible
*Ground Fault Circuit Interrupters (GCFI) within two feet of water supplies
*signs designating room exits
*emergency shower capable of providing a continuous flow of water
*emergency eye wash station which can be activated by the foot or forearm
*eye protection for every student and a means of sanitizing equipment
*emergency exhaust fans providing ventilation to the outside of the building
*master cut-off switches for gas, electric and compressed air. Switches must have permanently attached handles. Cut-off switches must be clearly labeled.
*an ABC fire extinguisher
*storage cabinets for flammable materials
-chemical spill control kit
-fume hood with a motor which is spark proof
-protective laboratory aprons made of flame retardant material
-signs which will alert potential hazardous conditions
-containers for broken glassware, flammables, corrosives, and waste. Containers should be labeled.

Students should wear safety goggles when performing dissections, heating, or while using acids and bases. Hair should always be tied back and objects should never be placed in the mouth. Food should not be consumed while in the laboratory. Hands should always be washed before and after laboratory experiments. In case of an accident, eye washes and showers should be used for eye contamination or a chemical spill that covers the student's body. Small chemical spills should only be contained and cleaned by the teacher. Kitty litter or a chemical spill kit should be used to clean spill. For large spills, the school administration and the local fire department should be notified. Biological spills should also be handled only by the teacher. Contamination with biological waste can be cleaned by using bleach when appropriate.

Accidents and injuries should always be reported to the school administration and local health facilities. The severity of the accident or injury will determine the course of action to pursue.

It is the responsibility of the teacher to provide a safe environment for their students. Proper supervision greatly reduces the risk of injury and a teacher should never leave a class for any reason without providing alternate supervision. After an accident, two factors are considered; **foreseeability** and **negligence**. Foreseeability is the anticipation that an event may occur under certain circumstances. Negligence is the failure to exercise ordinary or reasonable care. Safety procedures should be a part of the science curriculum and a well managed classroom is important to avoid potential lawsuits.

All laboratory solutions should be prepared as directed in the lab manual. Care should be taken to avoid contamination. All glassware should be rinsed thoroughly with distilled water before using and cleaned well after use. All solutions should be made with distilled water as tap water contains dissolved particles that may affect the results of an experiment. Unused solutions should be disposed of according to local disposal procedures.

The "Right to Know Law" covers science teachers who work with potentially hazardous chemicals. Briefly, the law states that employees must be informed of potentially toxic chemicals. An inventory must be made available if requested. The inventory must contain information about the hazards and properties of the chemicals. This inventory is to be checked against the "Substance List". Training must be provided on the safe handling and interpretation of the Material Safety Data Sheet.

The following chemicals are potential carcinogens and not allowed in school facilities: Acrylonitriel, Arsenic compounds, Asbestos, Bensidine, Benzene, Cadmium compounds, Chloroform, Chromium compounds, Ethylene oxide, Ortho-toluidine, Nickel powder, and Mercury.

Chemicals should not be stored on bench tops or heat sources. They should be stored in groups based on their reactivity with one another and in protective storage cabinets. All containers within the lab must be labeled. Suspect and known carcinogens must be labeled as such and segregated within trays to contain leaks and spills.

Chemical waste should be disposed of in properly labeled containers. Waste should be separated based on their reactivity with other chemicals.

Biological material should never be stored near food or water used for human consumption. All biological material should be appropriately labeled. All blood and body fluids should be put in a well-contained container with a secure lid to prevent leaking. All biological waste should be disposed of in biological hazardous waste bags.

Material safety data sheets are available for every chemical and biological substance. These are available directly from the company of acquisition or the internet. The manuals for equipment used in the lab should be read and understood before using them.

COMPETENCY 4.0 UNDERSTAND AND APPLY KNOWLEDGE OF THE INTERACTIONS AMONG SCIENCE, TECHNOLOGY, AND SOCIETY.

Skill 4.1 Recognize the historical and contemporary development of major scientific ideas and technological innovations.

Development of Modern Chemistry

Chemistry emerged from two ancient roots: **craft traditions** and **philosophy**. The oldest ceramic crafts (i.e., pottery) known are from roughly 10000 BC in Japan. **Metallurgical crafts** in Eurasia and Africa began to develop by trial and error around 4000-2500 BC resulting in the production of copper, bronze, iron, and steel tools. Other craft traditions in brewing, tanning, and dyeing led to many useful empirical ways to manipulate matter.

Ancient philosophers in Greece, India, China, and Japan speculated that all matter was composed of four or five elements. The Greeks thought that these were: fire, air, earth, and water. Indian philosophers and the Greek **Aristotle** also thought a fifth element—"aether" or "quintessence"—filled all of empty space. The Greek philosopher **Democritus** thought that matter was composed of indivisible and indestructible atoms. These concepts are now known as **classical elements** and **classical atomic theory**.

Before the emergence of the **scientific method**, attempts to understand matter relied on **alchemy**: a mixture of mysticism, best guesses, and supernatural explanations. Goals of alchemy were the transmutation of other metals into gold and the synthesis of an elixir to cure all diseases. Ancient Egyptian alchemists developed cement and glass. Chinese alchemists developed gunpowder in the 800s AD.

During the height of European alchemy in the 1300s, the philosopher **William of Occam** proposed the idea that when trying to explain a process or develop a theory, **the simplest explanation with the fewest variables is best**. This is known as **Occam's Razor**. European alchemy slowly developed into modern chemistry during the 1600s and 1700s. This began to occur after **Francis Bacon** and René **Descartes** described the scientific method in the early 1600s.

Robert **Boyle** was educated in alchemy in the mid-1600s, but he published a book called *The Skeptical Chemist* that attacked alchemy and advocated using the scientific method. He is sometimes called **the founder of modern chemistry** because of his emphasis on proving a theory before accepting it, but the birth of modern chemistry is usually attributed to Lavoisier. Boyle rejected the 4 classical elements and proposed the modern definition of an element. **Boyle's law** states that gas volume is proportional to the reciprocal of pressure.

Blaise **Pascal** in the mid-1600s determined the relationship between **pressure** and the height of a liquid in a **barometer**. He also helped to establish the scientific method. The SI unit of pressure is named after him.

Isaac **Newton** studied the nature of **light**, the laws of **gravity**, and the **laws of motion** around 1700. The SI unit of force is named after him.

Daniel **Bernoulli** proposed the **kinetic molecular theory** for gases in the early 1700s to explain the nature of heat and Boyle's Law. At that time, heat was thought to be related to the release of a substance called *phlogiston* from combustible materials

James **Watt** created an efficient **steam engine** in the 1760s-1780s. Later chemists and physicists would develop the theory behind this empirical engineering accomplishment. The SI unit of power is named after him.

Joseph **Priestley** studied various gases in the 1770s. He was the first to produce and drink **carbonated water**, and he was the first to isolate **oxygen** from air. Priestley thought oxygen was air with its normal phlogiston removed so it could burn more fuel and accept more phlogiston than natural air.

Antoine **Lavoisier** is called **the father of modern chemistry** because he performed **quantitative, controlled experiments**. He carefully weighed material before and after combustion to determine that burning objects gain weight. Lavoisier formulated the rule that **chemical reactions do not alter total mass** after finding that reactions in a closed container do not change weight. This disproved the plogiston theory, and he named Priestley's substance oxygen. He demonstrated that air and water were not elements. He defined an element as a substance that could not be broken down further. He published the first modern chemistry textbook, *Elementary Treatise of Chemistry*. Lavoisier was executed in the Reign of Terror at the height of the French Revolution.

Additional Gas Laws in the 1700s and 1800s
These contributions built on the foundation developed by Boyle in the 1600s.

Jacque **Charles** developed **Charles's law** in the late 1700s. This states that gas volume is proportional to absolute temperature.

William Henry developed the law stating that gas solubility in a liquid is proportional to the pressure of gas over the liquid. This is known as **Henry's Law**.

Joseph Louis **Gay-Lussac** developed the gas law stating that gas pressure is directly proportional to absolute temperature. He also determined that two volumes of hydrogen react with one of oxygen to produce water and that other reactions occurred with similar simple ratios. These observations led him to develop the **Law of Combining Volumes**.

Amedeo **Avogadro** developed the hypothesis that **equal volumes of different gases contain an equal numbers of molecules** if the gases are at the same temperature and pressure. The proportionality between volume and number of moles is called **Avagadro's Law**, and the number of molecules in a mole is called **Avagadro's Number**. Both were posthumously named in his honor.

Thomas **Graham** developed **Graham's Law** of effusion and diffusion in the 1830s. He is called the father of **colloid chemistry** .

Electricity and Magnetism in the 1700s and 1800s
Benjamin Franklin studied electricity in the mid-1700s. He developed the concept of **positive and negative electrical charges**. His most famous experiment showed that lightning is an electrical process.

Luigi **Galvani** discovered **bioelectricity**. In the late 1700s, he noticed that the legs of dead frogs twitched when they came into contact with an electrical source.

In the late 1700s, Charles Augustin **Coulomb** derived mathematical **equations for attraction and repulsion** between electrically charged objects.

Alessandro **Volta** built the first **battery** in 1800 permitting future research and applications to have a source of continuous electrical current available. The SI unit of electric potential difference is named after him.

André-Marie **Ampère** created a mathematical theory in the 1820s for magnetic fields and electric currents. The SI unit of electrical current is named after him.

Michael **Faraday** is best known for work in the 1820s and 1830s establishing that a moving magnetic field induces an electric potential. He built the first **dynamo** for electricity generation. He also discovered benzene, invented oxidation numbers, and popularized the terms *electrode*, *anode*, and *cathode*. The SI unit of electrical capacitance is named in his honor.

James Clerk **Maxwell** derived the **Maxwell Equations** in 1864. These expressions completely describe **electric and magnetic fields** and their interaction with matter. Also see Ludwig Boltzmann below for Maxwell's contribution to thermodynamics.

PHYSICS

Nineteenth Century Chemistry: Caloric Theory and Thermodynamics

Lavoisier proposed in the late 18th century that the heat generated by combustion was due to a weightless material substance called **caloric** that flowed from one place to another and was never destroyed.

In 1798, **Benjamin Thomson**, also known as **Count Rumford** measured the heat produced when cannon were bored underwater and concluded that caloric was not a conserved substance because heat could continue to be generated indefinitely by this process.

Sadi **Carnot** in the 1820s used caloric theory in developing theories for the **heat engine** to explain the engine already developed by Watt. Heat engines perform mechanical work by expanding and contracting a piston at two different temperatures.

In the 1820s, Robert **Brown** observed dust particles and particles in pollen grains moving in a random motion. This was later called **Brownian motion**.

Germain Henri **Hess** developed **Hess's Law** in 1840 after studying the heat required or emitted from reactions composed of several steps.

James Prescott **Joule** determined the equivalence of heat energy to mechanical work in the 1840s by carefully measuring the heat produced by friction. Joule attacked the caloric theory and played a major role in the acceptance of **kinetic molecular theory**. The SI unit of energy is named after him.

William Thomson, 1st Baron of Kelvin also called **Lord Kelvin** recognized the existence of **absolute temperature** in the 1840s and proposed the temperature scale named after him. He failed in an attempt to reconcile caloric theory with Joule's discovery and caloric theory began to fall out of favor.

Hermann von **Helmholtz** in the 1840s proposed that **energy is conserved** during physical and chemical processes, not heat as proposed in caloric theory

Rudolf **Clausius** in the 1860s introduced the concept of **entropy**.

In the 1870s, Ludwig **Boltzmann** generalized earlier work by Maxwell solving the **velocity or energy distribution among gas molecules**. The final diagram in **Skill 6.4** shows the Maxwell-Boltzmann distribution for kinetic energy at two temperatures. Maxwell's contribution to electromagnetism is described above.

Johannes **van der Waals** in the 1870s was the first to consider **intermolecular attractive forces** in modeling the behavior of liquids and non-ideal gases.

Francois Marie **Raoult** studied colligative properties in the 1870s. He developed **Raoult's Law** relating solute and solvent mole fraction to vapor pressure lowering.

Jacobus **van't Hoff** was the first to fully describe **stereoisomerism** in the 1870s. He later studied **colligative properties** and the impact of temperature on equilibria

Josiah Willard **Gibbs** studied thermodynamics and statistical mechanics in the 1870s. He formulated the concept now called **Gibbs free energy** that will determine whether or not a chemical process at constant pressure will spontaneously occur.

Henri Louis **Le Chatelier** described chemical **equilibrium** in the 1880s using **Le Chatelier's Principle**.

In the 1880s, Svante **Arrhenius** developed the idea of **activation energy**. He also described the dissociation of salts—including **acids and bases**—into ions. Before then, salts in solution were thought to exist as intact molecules and ions were mostly thought to exist as electrolysis products. Arrhenius also predicted that CO_2 emissions would lead to global warming.

In 1905, **Albert Einstein** created a **mathematical model of Brownian motion** based on the impact of water molecules on suspended particles. Kinetic molecular theory could now be observed under the microscope. Einstein's more famous later work in physics on **relativity** may be applied to chemistry by correlating the energy change of a chemical reaction with extremely small changes in the total mass of reactants and products.

<u>Nineteenth and Twentieth Century: Atomic Theory</u>

The contributions to atomic theory of John **Dalton**, **J. J. Thomson,** Max **Planck,** Ernest **Rutherford,** Niels **Bohr,** Louis **de Broglie,** Werner **Heisenberg,** and Erwin **Schrödinger** are discussed later in this book.

Wolfgang **Pauli** helped to develop quantum mechanics in the 1920s by forming the concept of spin and the **exclusion principle**.

Friedrich **Hund** determined a set of **rules to determine the ground state** of a multi-electron atom in the 1920s. One particular rule is called **Hund's Rule** in introductory chemistry courses.

Discovery and Synthesis: Nineteenth Century:

Humphry **Davy** used Volta's battery in the early 1800s for **electrolysis of salt solutions**. He synthesized several pure elements using electrolysis to generate non-spontaneous reactions.

Jöns Jakob **Berzelius** isolated several elements, but he is best known for inventing modern **chemical notation** by using one or two letters to represent elements in the early 1800s.

Friedrich **Wöhler** isolated several elements, but he is best known for the chemical **synthesis of an organic compound** in 1828 using the carbon in silver cyanide. Before Wöhler, many had believed that a transcendent "life-force" was needed to make the molecules of life.

Justus **von Liebig** studied the chemicals involved in agriculture in the 1840s. He has been called the **father of agricultural chemistry**.

Louis **Pasteur** studied **chirality** in the 1840s by separating a mixture of two chiral molecules. His greater contribution was in biology for discovering the germ theory of disease.

Henry **Bessemer** in the 1850s developed the **Bessemer Process** for mass producing steel by blowing air through molten iron to oxidize impurities.

Friedrich August **Kekulé** von Stradonitz studied the chemistry of carbon in the 1850s and 1860s. He proposed the **ring structure of benzene** and that carbon was tetravalent.

Anders Jonas **Ångström** was one of the founders of the science of spectroscopy. In the 1860s, he found hydrogen and other **elements in the spectrum of the sun**. A non-SI unit of length equal to 0.1 nm is named for him.

Alfred **Nobel** invented the explosive **dynamite** in the 1860s and continued to develop other explosives. In his will he used his fortune to establish the **Nobel Prizes**.

Dmitri **Mendeleev** developed the first modern **periodic table** in 1869.

Discovery and Synthesis: Turn of the 20th Century

William **Ramsay** and Lord **Rayleigh** (John William Strutt) isolated the **noble gases**.

Wilhelm Konrad **Röntgen** discovered **X-rays**.

Antoine Henri **Becquerel discovered radioactivity** using uranium salts.

Marie **Curie** named the property radioactivity and determined that it was **a property of atoms** that did not depend on which molecule contained the element.

Pierre and Marie **Curie** utilized the properties of radioactivity to **isolate radium** and other radioactive elements. Marie Curie was the first woman to receive a Nobel Prize and the first person to receive two. Her story continues to inspire. See http://nobelprize.org/physics/articles/curie/index.html for a biography.

Frederick **Soddy** and William **Ramsay** discovered that **radioactive decay can produce helium** (alpha particles).

Fritz **Haber** developed the **Haber Process** for synthesizing ammonia from hydrogen and nitrogen using an iron **catalyst**. Ammonia is still produced by this method to make fertilizers, textiles, and other products.

Robert Andrew **Millikan** determined the **charge of an electron** using an oil-drop experiment.

Discovery and Synthesis: 20th Century

Gilbert Newton **Lewis** described **covalent bonds** as sharing electrons in the 1910s and the **electron pair donor/acceptor theory of acids and bases** in the 1920s. Lewis dot structures (see **Skill 4.3**) and Lewis acids are named after him.

Johannes Nicolaus **Brønsted** and Thomas Martin **Lowry** simultaneously developed the **proton donor/acceptor theory of acids and bases** in the 1920s.

Irving **Langmuir** in the 1920s developed the science of **surface chemistry** to describe interactions at the interface of two phases. This field is important to heterogeneous catalysis.

Fritz **London** studied the electrical nature of chemical bonding in the 1920s. The weak intermolecular **London dispersion forces** are named after him.

Hans Wilhelm **Geiger** developed the **Geiger counter** for measuring ionizing radiation in the 1930s.

Wallace **Carothers** and his team first synthesized **organic polymers** (including neoprene, polyester and nylon) in the 1930s.

In the 1930s, Linus **Pauling** published his results on **the nature of the covalent bond**. Pauling electronegativity is named after him. In the 1950s, Pauling determined the α-helical structure of proteins.

Lise **Meitner** and Otto **Hahn** discovered **nuclear fission** in the 1930s.

Glenn Theodore **Seaborg** created and isolated several **elements larger than uranium** in the 1940s. Seaborg reorganized the periodic table to its current form.

James **Watson** and Francis **Crick** determined the double helix structure of DNA in the 1950s.

Neil **Bartlett** produced **compounds containing noble gases** in the 1960s, proving that they are not completely chemically inert.

Harold Kroto, Richard Smalley, and Robert Curl discovered the **buckyball C_{60}** in the 1980s.

Skill 4.2 **Demonstrate an understanding of the ways that science and technology affect people's everyday lives, societal values and systems, the environment, and new knowledge.**

Society as a whole impacts biological research. The pressure from the majority of society has led to these bans and restrictions on human cloning research. Human cloning has been restricted in the United States and many other countries. The U.S. legislature has banned the use of federal funds for the development of human cloning techniques. Some individual states have banned human cloning regardless of where the funds originate.

The demand for genetically modified crops by society and industry has steadily increased over the years. Genetic engineering in the agricultural field has led to improved crops for human use and consumption. Crops are genetically modified for increased growth and insect resistance because of the demand for larger and greater quantities of produce.

With advances in biotechnology come those in society who oppose it. Ethical questions come into play when discussing animal and human research. Does it need to be done? What are the effects on humans and animals? There are not right or wrong answers to these questions. There are governmental agencies in place to regulate the use of humans and animals for research.

Science and technology are often referred to as a "double-edged sword". Although advances in medicine have greatly improved the quality and length of life, certain moral and ethical controversies have arisen. Unforeseen environmental problems may result from technological advances. Advances in science have led to an improved economy through biotechnology as applied to agriculture, yet it has put our health care system at risk and has caused the cost of medical care to skyrocket. Society depends on science, yet is necessary that the public be scientifically literate and informed in order to allow potentially unethical procedures to occur. Especially vulnerable are the areas of genetic research and fertility. It is important for science teachers to stay abreast of current research and to involve students in critical thinking and ethics whenever possible.

Skill 4.3 Analyze the processes of scientific and technological breakthroughs and their effects on other fields of study, careers, and job markets.

Scientific and technological breakthroughs greatly influence other fields of study and the job market. All academic disciplines utilize computer and information technology to simplify research and information sharing. In addition, advances in science and technology influence the types of available jobs and the desired work skills. For example, machines and computers continue to replace unskilled laborers and computer and technological literacy is now a requirement for many jobs and careers. Finally, science and technology continue to change the very nature of careers. Because of science and technology's great influence on all areas of the economy, and the continuing scientific and technological breakthroughs, careers are far less stable than in past eras. Workers can thus expect to change jobs and companies much more often than in the past.

Skill 4.4 Analyze issues related to science and technology at the local, state, national, and global levels (e.g., environmental policies, genetic research).

A baby born today in the United States is expected to live 30 years longer on average than a baby born 100 years ago. A significant cause of this improvement is due to chemical technology. The manufacture and distribution of vaccines and antibiotics, an increase in understanding human nutritional needs, and the use of fertilizers in agriculture have all played a significant role in improving the length and the quality of human life. But the benefits of these technologies are almost always accompanied by problems and significant risks.

Nutrition: general

Chemical technology helps **keep foods fresh** longer and **alters the molecules** in food. The thermochemistry of refrigeration helps food last longer without altering it. Other technologies such as pasteurization, drying, salting, and the addition of preservatives all prevent microbial contamination by altering the nutritional content of food. **Preservatives** are substances added to food to prevent the growth of microorganisms and spoilage. For example, potassium and sodium **nitrites** and **nitrates** are often used as a preservative for root vegetables and processed meats.

Physical separation techniques such as milling, centrifugation, and pressing give flour, oils, and juices that are used as ingredients. **Chemical techniques** give particular fatty acids, amino acids, vitamins, and minerals that are used in nutritional supplements or to fortify processed foods. Some molecules in processed foods are removed intentionally or as an unintended consequence of processing. Other molecules are added for a wide variety of reasons such as improving taste or decreasing the cost of production.

Eating too much processed food over time has a long history of **causing harm** because of the substances it lacks or contains. Whole, fresh food usually has a better nutritional value. For example, in the late 1800s many infants in the US developed **scurvy (vitamin C deficiency)** by drinking heat-treated milk that controlled bacterial infections but destroyed vitamin C. Local production of food with minimal time-to-market and proper preparation seem to be a more sensible approach to food safety than long-distance transport of both the food and the means to protect it, but processed food will remain popular for the foreseeable future because it is usually **cheaper** to buy, more **profitable** to sell, and more **convenient** to obtain, store, prepare and use.

Nutrition: hydrogenation

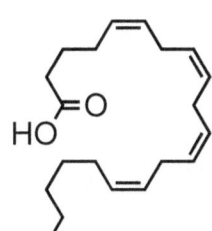

Hydrogenation uses a chemical reaction to convert unsaturated to saturated oils. Many plant oils are **polyunsaturated** with double bonds in the *cis-* form as shown at left. These molecules contain rigid bends in them. Complete hydrogenation creates a flexible straight-chain molecule that permits more area for London dispersion forces to form intermolecular bonds. The result is that **hydrogenation increases the melting point** of an oil. Semi-solid fats are preferred for baking because the final product has the right texture in the mouth. Unfortunately, saturated fats are less healthy than *cis-* unsaturated fats because they promote obesity and heart disease. Complete hydrogenation of the molecule above is shown here:

When the hydrogenation process does not fully saturate, it results in partially **hydrogenated** oil. Partial hydrogenation often creates a semi-solid fat in cases where complete hydrogenation would create a fat that is fully solid.

However, **partial hydrogenation** of cis-polyunsaturated fats results in a **random isomerization** creating a **mixture of *cis*- and *trans*-** forms. In the structure below, the molecule at the top of the page has been partially hydrogenated, resulting in the saturation of two double bonds. One of the two remaining *cis*-bonds has been isomerized to a *trans*- form:

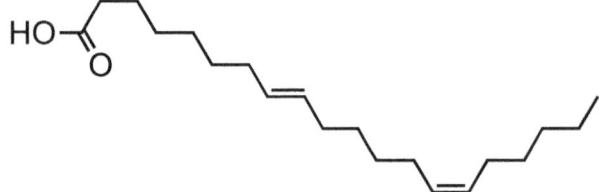

Trans-fatty acids have a slight kink in them compared to *cis*- forms, and they rarely occur in the food found in nature. Campaigns against saturated fat in the 1980s led to the increased use of partially hydrogenated oils. The health benefits of **monounsaturated fat** were promoted, but labels made **no distinction between *cis*- and *trans*-** forms. As a result, there has been an increase in consumption of *trans* fat. Unfortunately, it is now known that ***trans* fat is even worse for the body than saturated fat**. Some nations have completely banned the use of partially hydrogenated oils. Food labels in the United States are currently (as of 2006) required to list **total, saturated, and *trans*-fat content**. Fatty acids with one or more *trans* nonconjugated double bonds are labeled as *trans* fat under this rule.

Nutrition: irradiation
Food may also be sterilized and preserved by **food irradiation**. **Gamma rays** from a sealed source of ^{60}Co or ^{137}Cs are used to **kill microorganisms** in over 40 countries. This process is less expensive than refrigeration, canning, or additives, and it **does not make food radioactive**.

Opponents of irradiation fear the risks involved in the **transport and use of nuclear materials** to build the facilities and to maintain them. A potential health risk in the food itself is the possibility that the radiation required to kill organisms may alter a biological molecule to produce a harmful by-product, but no evidence has been found of such a toxin. Another concern is that irradiation will lead to a permissive attitude about safe food-handling procedures that can lead to other types of contamination. Food irradiation is still under study in the United States to conclusively prove its safety, particularly for meats

Industry

The *chemical industry* usually refers to **industries that manufacture chemicals**. But it could be argued that the impact of those chemicals and of the ability to alter matter by chemical technology has **created tools that have improved the industrial production of nearly every substance**. Some divisions of the chemical industry are: petrochemicals (chemicals from petroleum), oleochemicals (chemicals from biological oils and fats), agrochemicals (chemicals for agriculture), pharmaceuticals, polymers, and paints. Chemical technology has impacted everything from testing and maintaining a clean water supply to the plastics used in cell phones.

These industries make up an entire **sector of the economy**. In the United States, about 900,000 people are employed in the chemical industry, and sales totaled about $500 billion in 2004 (http://pubs.acs.org/cen/coverstory/83/8302wcousa.html).

Chemical engineering is the **application of chemistry** along with mathematics and economics to the process of converting raw materials or chemicals into more useful forms. In the development of a new chemical, a chemist typically discovers or **designs the compound** and synthesizes it for the first time. A chemical engineer will then typically take that information and the **design the process** to manufacture the chemical in the amounts required. The individual processes used by chemical engineers (such as distillation or solvent extraction) are called **unit operations**. All chemical engineers are trained in process design, but most work in a variety of other disciplines.

Medicine

In the field of **medicinal chemistry**, scientists identify, synthesize, develop, and study chemicals to use for diagnostic tools and pharmaceuticals. **Pharmacology** is the study of how chemical substances interact with living systems. As biological knowledge has increased, the biochemical causes of many diseases have been determined and the field of pharmacology has grown tremendously.

Antibiotics are organic chemicals to **kill or slow the growth of bacteria**. Before antibiotics were available, infections were often treated with moderate levels of poisons like strychnine or arsenic. Antibiotics **target the disease without harming the patient**, and they have saved millions of lives. Unfortunately, some bacteria in **antibiotic resistant strains** have developed defenses against these chemicals since they first became available over 60 years ago. New antibiotics are developed every year in an effort to continue the suppression of bacterial diseases.

Biotechnology uses living organisms—often cells in a **fermentation tank** or **bioreactor**—to create useful molecules. The oldest examples are the use of yeast to make bread and beer. Since 1980, **genetic engineering** has been used to design **recombinant DNA** in order to produce **human protein** molecules in bioreactors using non-human cells. These molecules fight diseases by elevating the level of proteins made naturally by the human body or by providing proteins that are missing due to genetic disorders. Most tools in biotechnology originated from chemical technology.

Agriculture

Plants, like humans and animals, need adequate water, protection from disease, and certain chemical compounds to grow. Soil fertility where no agriculture occurs is maintained at an even level when the waste products derived from plants are returned to the soil. Human agriculture prevents this from occurring. **Fertilizers** are materials given to plants to **promote growth**. **Nitrogen, phosphorus, and potassium** are the most important elements in fertilizers.

There are many types of fertilizer, including natural materials like manure, seaweed, compost, and minerals from deposits. Plants that are able to utilize nitrogen from the atmosphere may also be grown some seasons so their nitrogen is added to the soil. Many of these technologies are thousands of years old. In the 1800s, studies by **von Liebig** and others resulted in the worldwide transport of a few minerals and by-products of the steel industry as fertilizer.

The major breakthrough in the use of fertilizers from chemical processes occurred with the development of the **Haber process for ammonia production** in 1910:

$N_2(g) + 3H_2(g) \rightleftharpoons 2NH_3(g)$ over Fe catalyst.

Millions of tons of ammonia are used worldwide each year to supply crops with nitrogen. Ammonia is either added to irrigation water or injected directly into the ground. Many other nitrogen fertilizers are synthesized from ammonia. Phosphorus in fertilizers originates from phosphate (PO_4^{3-}) in rock deposits. Potassium in fertilizers comes from evaporated ancient seabeds in the form of potassium oxide (K_2O).

Pesticides are used to control or kill organisms that compete with humans for food, spread disease, or are considered a nuisance. Herbicides are pesticides that attack weeds; insecticides attack insects; fungicides attack molds and other fungus. Sulfur was used as a fungicide in ancient times. The development and use of new pesticides has exploded over the last 60 years, but these pesticides are often poisonous to humans.

Farming designed to **maximize productivity** is called **intensive agriculture**. These methods of fertilizer and pesticide use in combination with other farming techniques decreased the number of farm laborers needed and gave a growing world population enough to eat in the last 50 years. Intensification of agriculture in developing countries is known as the **green revolution**. These techniques were credited with saving a billion people from starvation in India and Pakistan.

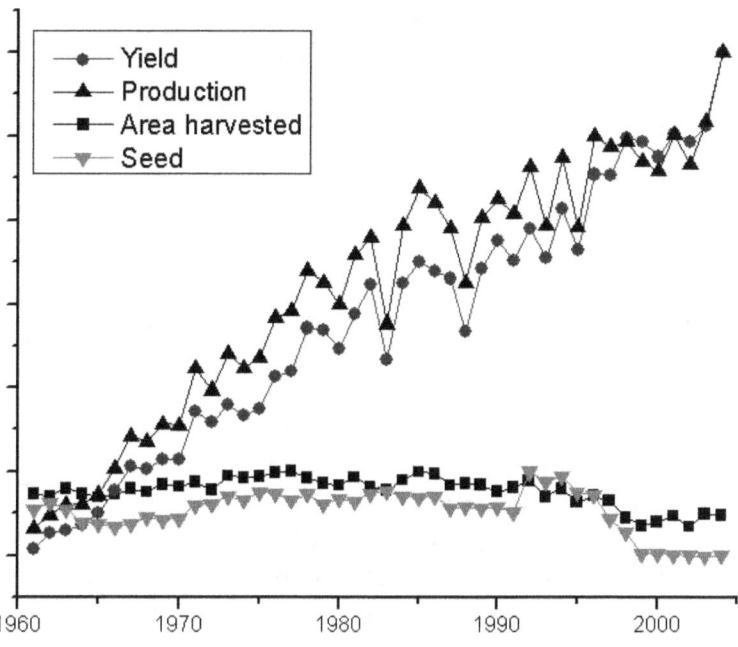

Total world production of coarse grain, 1961-2004

Source: Food and Agriculture Organization of the United Nations (http://faostat.fao.org/)

The **insecticide DDT** was widely used in the 1940s and 1950s and is responsible for **eradicating malaria from Europe and North America**. It quickly became the most widely used pesticide in the world. In the 1960s, some claimed that DDT was preventing fish-eating birds from reproducing and that it was causing birth defects in humans. DDT is now banned in many countries, but it is still used in developing nations to prevent diseases carried by insects. Unfortunately, its use in agriculture has often led to resistant mosquito strains that have hindered its effectiveness to prevent diseases.

The **herbicide *Roundup*** kills all natural plants it encounters. It began to be used in the 1990s in combination with **genetically engineered crops** that include a gene intended to make the crop (and only the crop) resistant to the herbicide. This combination of chemical and genetic technology has been an economic success but it has raised many concerns about potential problems in the future.

Environment

Many chemical technologies that save lives or improve the quality of life in the short term have had a negative impact on the environment. Chemical **pollution** is often divided into gas, liquid, and solid waste materials. Additional technologies often exist remediate these effects and improve pollution.

Most scientists believe the emission of **greenhouse gases** has already led to **global warming** due to an increase in **the greenhouse effect**. The greenhouse effect occurs when these gases in the atmosphere warm the planet by **absorbing heat** to prevent it from escaping into space. This is similar—but not identical—to what occurs in greenhouse buildings. Greenhouse buildings warm an interior space by preventing mixing with colder gases outside. Many greenhouse gases such as water vapor occur naturally and are important for life to exist on Earth. Human production of **carbon dioxide** from combustion of fossil fuels has increased the concentration of this important greenhouse gas to its highest value since millions of years ago. The precise impact of these changes in the atmosphere is difficult to predict and is a topic of international concern and political debate.

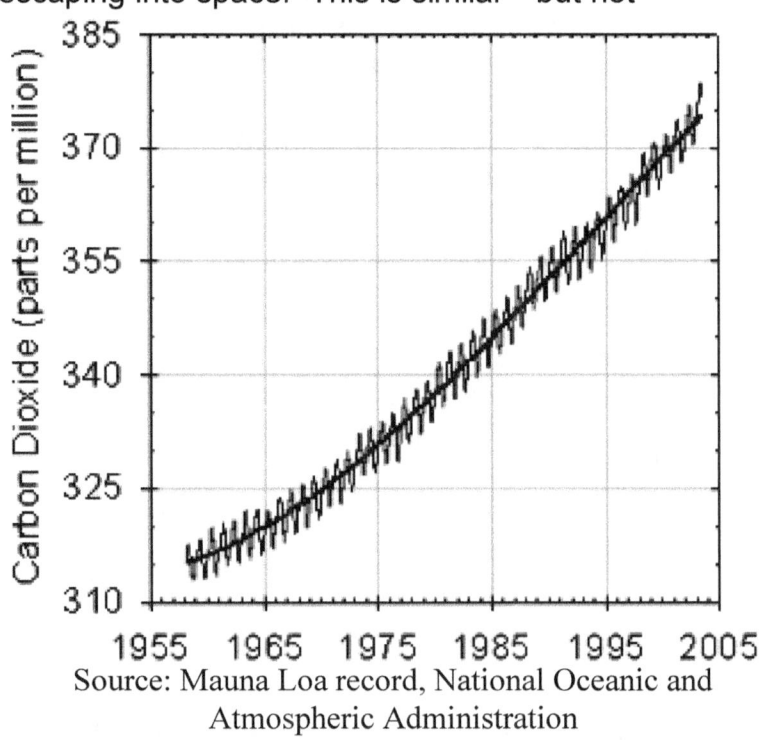
Source: Mauna Loa record, National Oceanic and Atmospheric Administration

Rain with a pH less than 5.6 is known as **acid rain**. Acid rain is caused by burning fossil fuels (especially coal) and by fertilizers used in intensive agriculture. These activities emit sulfur and nitrogen in gas compounds that are converted to sulfur oxides and nitrogen oxides. These in turn create sulfuric acid and nitric acid in rain. Acid rain may also be created from gases emitted by volcanoes and other natural sources. Acid rain harms fish and trees and triggers the release metal ions from minerals into water that can harm people. The problem of acid rain in the United States has been addressed in recent decades by the use of **scrubbers** in coal burning power plants and **catalytic converters** in vehicles.

Ozone is O_3. The **ozone layer** is a region of the stratosphere that contains higher concentrations of ozone than other parts of the atmosphere. The ozone layer is important for human health because it **blocks ultraviolet radiation** from the sun, and this helps to protect us from skin cancer. Research in the 1970s revealed that several gases used for refrigeration and other purposes were depleting the ozone layer. Many of these ozone-destroying molecules are short alkyl halides known as **chlorofluorocarbons** or **CFCs**. CCl_3F is one example. The widespread use of ozone-destroying gases was banned by an international agreement in the early 1990s. Other substances are used in their place such as CF_3CH_2F, a hydrofluorocarbon. Since that time the concentration of ozone-depleting gases in the atmosphere has been declining and **the rate of ozone destruction has been decreasing**. Many see this improvement as the most important positive example of international cooperation in helping the environment. The story of these new refrigerants is found at http://www.chemcases.com/fluoro/index.htm.

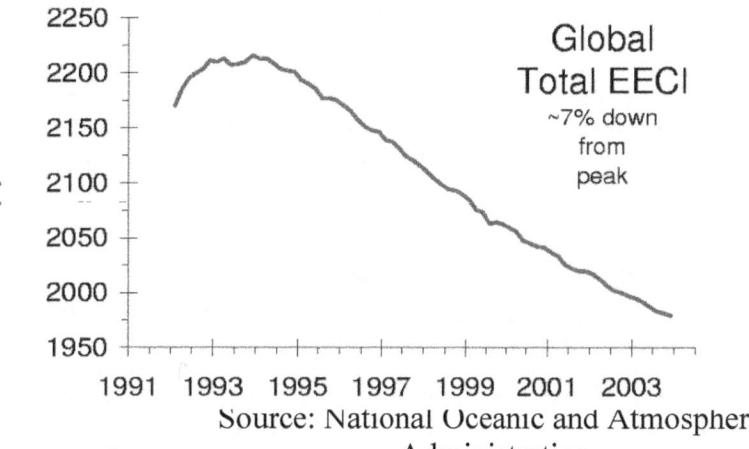

Source: National Oceanic and Atmospheric Administration

Skill 4.5 Evaluate the credibility of scientific claims made in various forums (e.g., the media, public debates, advertising).

Because people often attempt to use scientific evidence in support of political or personal agendas, the ability to evaluate the credibility of scientific claims is a necessary skill in today's society. In evaluating scientific claims made in the media, public debates, and advertising, one should follow several guidelines.

First, scientific, peer-reviewed journals are the most accepted source for information on scientific experiments and studies. One should carefully scrutinize any claim that does not reference peer-reviewed literature.

Second, the media and those with an agenda to advance (advertisers, debaters, etc.) often overemphasize the certainty and importance of experimental results. One should question any scientific claim that sounds fantastical or overly certain.

Finally, knowledge of experimental design and the scientific method is important in evaluating the credibility of studies. For example, one should look for the inclusion of control groups and the presence of data to support the given conclusions.

TEACHER CERTIFICATION STUDY GUIDE

COMPETENCY 5.0 UNDERSTAND AND APPLY KNOWLEDGE OF THE MAJOR UNIFYING CONCEPTS OF ALL SCIENCES AND HOW THESE CONCEPTS RELATE TO OTHER DISCIPLINES.

Skill 5.1 Identify the major unifying concepts of the sciences (e.g., systems, order, and organization; constancy, change, and measurement) and their applications in real-life situations.

The following are the concepts and processes generally recognized as common to all scientific disciplines:

- Systems, order, and organization
- Evidence, models, and explanation
- Constancy, change, and measurement
- Evolution and equilibrium
- Form and function

Because the natural world is so complex, the study of science involves the **organization** of items into smaller groups based on interaction or interdependence. These groups are called **systems**. Examples of organization are the periodic table of elements and the five-kingdom classification scheme for living organisms. Examples of systems are the solar system, cardiovascular system, Newton's laws of force and motion, and the laws of conservation.

Order refers to the behavior and measurability of organisms and events in nature. The arrangement of planets in the solar system and the life cycle of bacterial cells are examples of order.

Scientists use **evidence** and **models** to form **explanations** of natural events. Models are miniaturized representations of a larger event or system. Evidence is anything that furnishes proof.

Constancy and **change** describe the observable properties of natural organisms and events. Scientists use different systems of **measurement** to observe change and constancy. For example, the freezing and melting points of given substances and the speed of sound are constant under constant conditions. Growth, decay, and erosion are all examples of natural change.

Evolution is the process of change over a long period of time. While biological evolution is the most common example, one can also classify technological advancement, changes in the universe, and changes in the environment as evolution.

Equilibrium is the state of balance between opposing forces of change. Homeostasis and ecological balance are examples of equilibrium.

Form and **function** are properties of organisms and systems that are closely related. The function of an object usually dictates its form and the form of an object usually facilitates its function. For example, the form of the heart (e.g. muscle, valves) allows it to perform its function of circulating blood through the body.

Skill 5.2   Recognize connections within and among the traditional scientific disciplines.

Because biology is the study of living things, we can easily apply the knowledge of biology to daily life and personal decision-making. For example, biology greatly influences the health decisions humans make everyday. Other areas of daily life where biology affects decision-making are parenting, interpersonal relationships, family planning, and consumer spending.

What foods to eat, when and how to exercise, and how often to bathe are just three of the many decisions we make everyday that are based on our knowledge of science.

This is true for chemistry as well. The chair you sit in to read this manual is made of carbon, and the shampoo you used this morning is a combination of useful chemicals, probably of a synthetic nature. Science is everywhere!

Skill 5.3   Apply fundamental mathematical language, knowledge, and skills at the level of algebra and statistics in scientific contexts.

The knowledge and use of basic mathematical concepts and skills is a necessary aspect of scientific study. Science depends on data and the manipulation of data requires knowledge of mathematics. Understanding of basic statistics, graphs and charts, and algebra are of particular importance. Scientists must be able to understand and apply the statistical concepts of mean, median, mode, and range to sets of scientific data. In addition, scientists must be able to represent data graphically and interpret graphs and tables. Finally, scientists often use basic algebra to solve scientific problems and design experiments. For example, the substitution of variables is a common strategy in experiment design. Also, the ability to determine the equation of a curve is valuable in data manipulation, experimentation, and prediction.

Modern science uses a number of disciplines to understand it better. Statistics is one of those subjects that is absolutely essential for science.

Mean: Mean is the mathematical average of all the items. To calculate the mean, all the items must be added up and divided by the number of items. This is also called the arithmetic mean or more commonly as the "average".

Median: The median depends on whether the number of items is odd or even. If the number is odd, then the median is the value of the item in the middle. This is the value that denotes that the number of items having higher or equal value to that is same as the number of items having equal or lesser value than that. If the number o the items is even, the median is the average of the two items in the middle, such that the number of items having values higher or equal to it is same as the number o items having values equal or less than that.

Mode: Mode is the value o the item that occurs the most often, if there are not many items. Bimodal is a situation where there are two items with equal frequency.

Range: Range is the difference between the maximum and minimum values. The range is the difference between two extreme points on the distribution curve.

Scientists use mathematical tools and equations to model and solve scientific problems. Solving scientific problems often involves the use of quadratic, trigonometric, exponential, and logarithmic functions.

Quadratic equations take the standard form $ax^2 + bx + c = 0$. The most appropriate method of solving quadratic equations in scientific problems is the use of the quadratic formula. The quadratic formula produces the solutions of a standard form quadratic equation.

$$x = \frac{-b \pm \sqrt{b^2 - 4ac}}{2a}$$

One common application of quadratic equations is the description of biochemical reaction equilibriums. Consider the following problem.

> Example 1
>
> 80.0 g of ethanoic acid (MW = 60g) reacts with 85.0 g of ethanol (MW = 46g) until equilibrium. The equilibrium constant is 4.00. Determine the amounts of ethyl acetate and water produced at equilibrium.
>
> $$CH_3COOH + CH_3CH_2OH = CH_3CO_2C_2H_5 + H_2O$$
>
> The equilibrium constant, K, describes equilibrium of the reaction, relating the concentrations of products to reactants.
>
> $$K = \frac{[CH_3CO_2C_2H_5][H_2O]}{[CH_3CO_2H][CH_3CH_2OH]} = 4.00$$

The equilibrium values of reactants and products are listed in the following table.

	CH_3COOH	CH_3CH_2OH	$CH_3CO_2C_2H_5$	H_2O
Initial	80/60 = 1.33 mol	85/46 = 1.85 mol	0	0
Equilibrium	1.33 − x	1.85 − x	x	x

Thus, $K = \dfrac{[x][x]}{[1.33-x][1.85-x]} = \dfrac{x^2}{2.46-3.18x+x^2} = 4.00$.

Rearrange the equation to produce a standard form quadratic equation.

$$\dfrac{x^2}{2.46-3.18x+x^2} = 4.00$$

$$x^2 = 4.00(2.46-3.18x+x^2) = 9.84 - 12.72x + 4x^2$$

$$0 = 3x^2 - 12.72x + 9.84$$

Use the quadratic formula to solve for x.

$$x = \dfrac{-(-12.72) \pm \sqrt{(-12.72)^2 - 4(3)(9.84)}}{2(3)} = 3.22 \text{ or } 1.02$$

3.22 is not an appropriate answer, because we started with only 3.18 moles of reactants. Thus, the amount of each product produced at equilibrium is 1.02 moles.

Scientists use trigonometric functions to define angles and lengths. For example, field biologists can use trigonometric functions to estimate distances and directions. The basic trigonometric functions are sine, cosine, and tangent. Consider the following triangle describing these relationships.

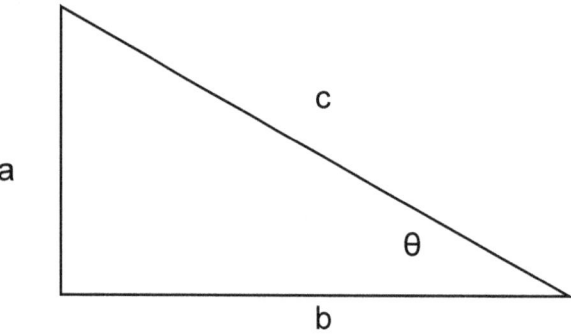

$$\sin \theta = \dfrac{a}{c}, \cos \theta = \dfrac{b}{c}, \tan \theta = \dfrac{a}{b}$$

Exponential functions are useful in modeling many scientific phenomena. For example, scientists use exponential functions to describe bacterial growth and radioactive decay. The general form of exponential equations is $f(x) = Ca^x$ (C is a constant). Consider the following problem involving bacterial growth.

Example 2

Determine the number of bacteria present in a culture inoculated with a single bacterium after 24 hours if the bacterial population doubles every 2 hours. Use $N(t) = N_0 e^{kt}$ as a model of bacterial growth where N(t) is the size of the population at time t, N_0 is the initial population size, and k is the growth constant.

We must first determine the growth constant, k. At t = 2, the size of the population doubles from 1 to 2. Thus, we substitute and solve for k.

$$2 = 1(e^{2k})$$

$\ln 2 = \ln e^{2k}$ Take the natural log of each side.

$\ln 2 = 2k(\ln e) = 2k$ $\ln e = 1$

$k = \dfrac{\ln 2}{2}$ Solve for k.

The population size at t = 24 is

$$N(24) = e^{(\frac{\ln 2}{2})24} = e^{12 \ln 2} = 4096.$$

Finally, logarithmic functions have many applications to science. One simple example of a logarithmic application is the pH scale. Scientists define pH as follows.

    pH = - $\log_{10}$ [H+], where [H+] is the concentration of hydrogen ions

Thus, we can determine the pH of a solution with a [H+] value of 0.0005 mol/L by using the logarithmic formula.

    pH = - $\log_{10}$ [0.0005] = 3.3

Skill 5.4 **Recognize the fundamental relationships among the natural sciences and the social sciences.**

The fundamental relationship between the natural and social sciences is the use of the scientific method and the rigorous standards of proof that both disciplines require. This emphasis on organization and evidence separates the sciences from the arts and humanities. Natural science, particularly biology, is closely related to social science, the study of human behavior. Biological and environmental factors often dictate human behavior and accurate assessment of behavior requires a sound understanding of biological factors.

TEACHER CERTIFICATION STUDY GUIDE

SUBAREA II.   LIFE SCIENCE

COMPETENCY 6.0   UNDERSTAND AND APPLY KNOWLEDGE OF CELL STRUCTURE AND FUNCTION.

Skill 6.1   Compare and contrast the structures of viruses and prokaryotic and eukaryotic cells.

The cell is the basic unit of all living things. There are three types of cells. They are prokaryotes, eukaryotes, and archaea. Archaea have some similarities with prokaryotes, but are as distantly related to prokaryotes as prokaryotes are to eukaryotes.

PROKARYOTES

Prokaryotes consist only of bacteria and cyanobacteria (formerly known as blue-green algae). The classification of prokaryotes is in the diagram below.

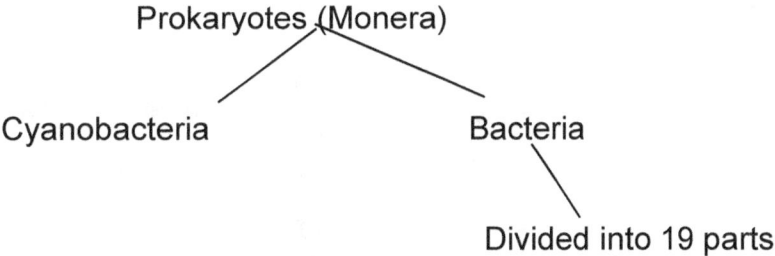

These cells have no defined nucleus or nuclear membrane. The DNA, RNA, and ribosomes float freely within the cell. The cytoplasm has a single chromosome condensed to form a **nucleoid**. Prokaryotes have a thick cell wall made up of amino sugars (glycoproteins). This is for protection, to give the cell shape, and to keep the cell from bursting. It is the **cell wall** of bacteria that is targeted by the antibiotic penicillin. Penicillin works by disrupting the cell wall, thus killing the cell.

The cell wall surrounds the **cell membrane** (plasma membrane). The cell membrane consists of a lipid bilayer that controls the passage of molecules in and out of the cell. Some prokaryotes have a capsule made of polysaccharides that surrounds the cell wall for extra protection from higher organisms.

Many bacterial cells have appendages used for movement called **flagella**. Some cells also have **pili**, which are a protein strand used for attachment of the bacteria. Pili may also be used for sexual conjugation (where the DNA from one bacterial cell is transferred to another bacterial cell).

Prokaryotes are the most numerous and widespread organisms on earth. Bacteria were most likely the first cells and date back in the fossil record to 3.5 billion years ago. Their ability to adapt to the environment allows them to thrive in a wide variety of habitats.

EUKARYOTES

Eukaryotic cells are found in protists, fungi, plants, and animals. Most eukaryotic cells are larger than prokaryotic cells. They contain many organelles, which are membrane bound areas for specific functions. Their cytoplasm contains a cytoskeleton which provides a protein framework for the cell. The cytoplasm also supports the organelles and contains the ions and molecules necessary for cell function. The cytoplasm is contained by the plasma membrane. The plasma membrane allows molecules to pass in and out of the cell. The membrane can bud inward to engulf outside material in a process called endocytosis. Exocytosis is a secretory mechanism, the reverse of endocytosis. The most significant differentiation between prokaryotes and eukaryotes is that eukaryotes have a **nucleus**.

VIRUSES

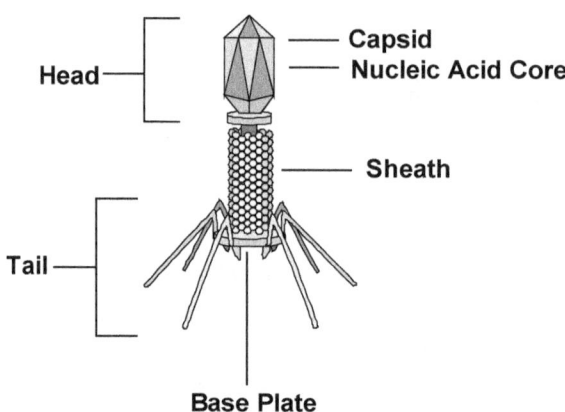

Bacteriophage

All viruses have a head or protein capsid that contains genetic material. This material is encoded in the nucleic acid and can be DNA, RNA, or even a limited number of enzymes. Some viruses also have a protein tail region. The tail aids in binding to the surface of the host cell and penetrating the surface of the host in order to introduce the virus's genetic material.

Other examples of viruses and their structures:

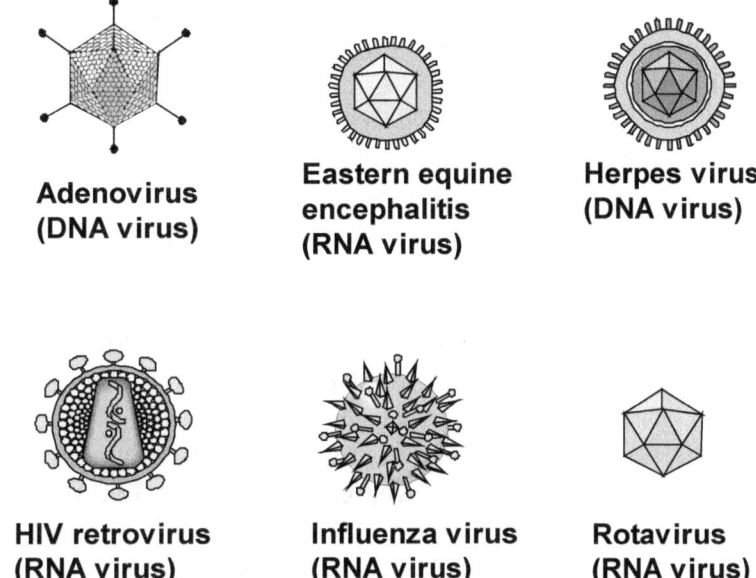

Skill 6.2 Identify the structures and functions of cellular organelles.

The **nucleus** is the brain of the cell that contains all of the cell's genetic information. The chromosomes consist of chromatin, which is a complex of DNA and proteins. The chromosomes are tightly coiled to conserve space while providing a large surface area. The nucleus is the site of transcription of the DNA into RNA. The **nucleolus** is where ribosomes are made. There is at least one of these dark-staining bodies inside the nucleus of most eukaryotes. The nuclear envelope is two membranes separated by a narrow space. The envelope contains many pores that let RNA out of the nucleus.

Ribosomes are the site for protein synthesis. Ribosomes may be free floating in the cytoplasm or attached to the endoplasmic reticulum. There may be up to a half a million ribosomes in a cell, depending on how much protein is made by the cell.

The **endoplasmic reticulum** (ER) is folded and provides a large surface area. It is the "roadway" of the cell and allows for transport of materials through and out of the cell. There are two types of ER. Smooth endoplasmic reticulum contains no ribosomes on their surface. This is the site of lipid synthesis. Rough endoplasmic reticulum has ribosomes on their surface. They aid in the synthesis of proteins that are membrane bound or destined for secretion.

Many of the products made in the ER proceed on to the Golgi apparatus. The **Golgi apparatus** functions to sort, modify, and package molecules that are made in the other parts of the cell (like the ER). These molecules are either sent out of the cell or to other organelles within the cell. The Golgi apparatus is a stacked structure to increase the surface area.

Lysosomes are found mainly in animal cells. These contain digestive enzymes that break down food, substances not needed, viruses, damaged cell components and eventually the cell itself. It is believed that lysosomes are responsible for the aging process.

Mitochondria are large organelles that are the site of cellular respiration, where ATP is made to supply energy to the cell. Muscle cells have many mitochondria because they use a great deal of energy. Mitochondria have their own DNA, RNA, and ribosomes and are capable of reproducing by binary fission if there is a greater demand for additional energy. Mitochondria have two membranes: a smooth outer membrane and a folded inner membrane. The folds inside the mitochondria are called cristae. They provide a large surface area for cellular respiration to occur.

Plastids are found only in photosynthetic organisms. They are similar to the mitochondira due to the double membrane structure. They also have their own DNA, RNA, and ribosomes and can reproduce if the need for the increased capture of sunlight becomes necessary. There are several types of plastids. **Chloroplasts** are the sight of photosynthesis. The stroma is the chloroplast's inner membrane space. The stoma encloses sacs called thylakoids that contain the photosynthetic pigment chlorophyll. The chlorophyll traps sunlight inside the thylakoid to generate ATP which is used in the stroma to produce carbohydrates and other products. The **chromoplasts** make and store yellow and orange pigments. They provide color to leaves, flowers, and fruits. The **amyloplasts** store starch and are used as a food reserve. They are abundant in roots like potatoes.

The Endosymbiotic Theory states that mitochondria and chloroplasts were once free living and possibly evolved from prokaryotic cells. At some point in our evolutionary history, they entered the eukaryotic cell and maintained a symbiotic relationship with the cell, with both the cell and organelle benefiting from the relationship. The fact that they both have their own DNA, RNA, ribosomes, and are capable of reproduction helps to confirm this theory.

Found in plant cells only, the **cell wall** is composed of cellulose and fibers. It is thick enough for support and protection, yet porous enough to allow water and dissolved substances to enter. **Vacuoles** are found mostly in plant cells. They hold stored food and pigments. Their large size allows them to fill with water in order to provide turgor pressure. Lack of turgor pressure causes a plant to wilt.

The **cytoskeleton**, found in both animal and plant cells, is composed of protein filaments attached to the plasma membrane and organelles. They provide a framework for the cell and aid in cell movement. They constantly change shape and move about. Three types of fibers make up the cytoskeleton:

1. **Microtubules** – the largest of the three, they make up cilia and flagella for locomotion. Some examples are sperm cells, cilia that line the fallopian tubes and tracheal cilia. Centrioles are also composed of microtubules. They aid in cell division to form the spindle fibers that pull the cell apart into two new cells. Centrioles are not found in the cells of higher plants.

2. **Intermediate filaments** – intermediate in size, they are smaller than microtubules but larger than microfilaments. They help the cell to keep its shape.

3. **Microfilaments** – smallest of the three, they are made of actin and small amounts of myosin (like in muscle tissue). They function in cell movement like cytoplasmic streaming, endocytosis, and ameboid movement. This structure pinches the two cells apart after cell division, forming two new cells.

The following is a diagram of a generalized animal cell.

ARCHAEA

There are three kinds of organisms with archaea cells: **methanogens** are obligate anaerobes that produce methane, **halobacteria** can live only in concentrated brine solutions, and **thermoacidophiles** can only live in acidic hot springs.

Skill 6.3   Describe the processes of the cell cycle.

The purpose of cell division is to provide growth and repair in body (somatic) cells and to replenish or create sex cells for reproduction. There are two forms of cell division. **Mitosis** is the division of somatic cells and **meiosis** is the division of sex cells (eggs and sperm).

Mitosis is divided into two parts: the **mitotic (M) phase** and **interphase**. In the mitotic phase, mitosis and cytokinesis divide the nucleus and cytoplasm, respectively. This phase is the shortest phase of the cell cycle. Interphase is the stage where the cell grows and copies the chromosomes in preparation for the mitotic phase. Interphase occurs in three stages of growth: **G1** (growth) period is when the cell is growing and metabolizing, the **S** period (synthesis) is where new DNA is being made and the **G2** phase (growth) is where new proteins and organelles are being made to prepare for cell division.

The mitotic phase is a continuum of change, although it is described as occurring in five stages: prophase, prometaphase, metaphase, anaphase, and telophase. During **prophase**, the cell proceeds through the following steps continuously, with no stopping. The chromatin condenses to become visible chromosomes. The nucleolus disappears and the nuclear membrane breaks apart. Mitotic spindles form that will eventually pull the chromosomes apart. They are composed of microtubules. The cytoskeleton breaks down and the spindles are pushed to the poles or opposite ends of the cell by the action of centrioles. During **prometaphase**, the nuclear membrane fragments and allows the spindle microtubules to interact with the chromosomes. Kinetochore fibers attach to the chromosomes at the centromere region. (Sometimes prometaphase is grouped with metaphase). When the centrosomes are at opposite ends of the cell, the division is in **metaphase**. The centromeres of all the chromosomes are aligned with one another. During **anaphase**, the centromeres split in half and homologous chromosomes separate. The chromosomes are pulled to the poles of the cell, with identical sets at either end. The last stage of mitosis is **telophase**. Here, two nuclei form with a full set of DNA that is identical to the parent cell. The nucleoli become visible and the nuclear membrane reassembles. A cell plate is seen in plant cells, whereas a cleavage furrow is formed in animal cells. The cell is pinched into two cells. Cytokinesis, or division of the cytoplasm and organelles, occurs.

Below is a diagram of mitosis.

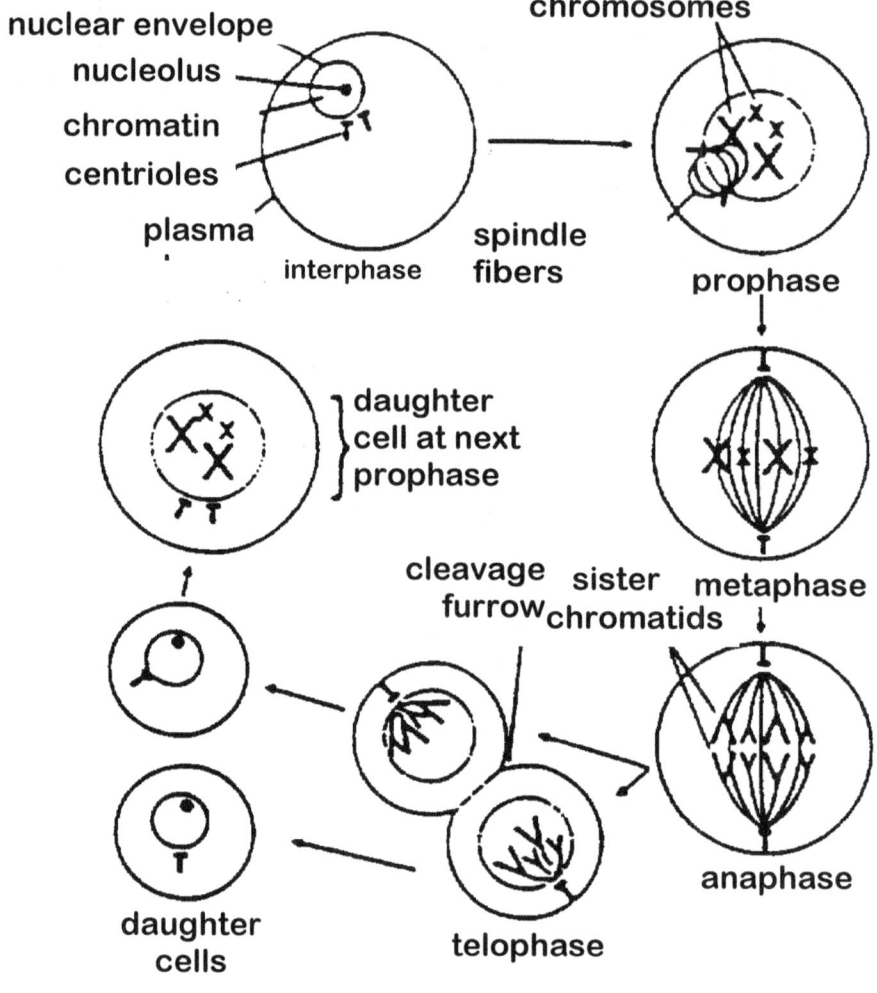

Mitosis in an Animal Cell

Meiosis is similar to mitosis, but there are two consecutive cell divisions, meiosis I and meiosis II in order to reduce the chromosome number by one half. This way, when the sperm and egg join during fertilization, the haploid number is reached.

Similar to mitosis, meiosis is preceded by an interphase during which the chromosome replicates. The steps of meiosis are as follows:

1. **Prophase I** – the replicated chromosomes condense and pair with homologues in a process called synapsis. This forms a tetrad. Crossing over, the exchange of genetic material between homologues to further increase diversity, occurs during prophase I.
2. **Metaphase I** – the homologous pairs attach to spindle fibers after lining up in the middle of the cell.
3. **Anaphase I** – the sister chromatids remain joined and move to the poles of the cell.
4. **Telophase I** – the homologous chromosome pairs continue to separate. Each pole now has a haploid chromosome set. Telophase I occurs simultaneously with cytokinesis. In animal cells, cleavage furrows form and cell plate appear in plant cells.
5. **Prophase II** – a spindle apparatus forms and the chromosomes condense.
6. **Metaphase II** – sister chromatids line up in center of cell. The centromeres divide and the sister chromatids begin to separate.
7. **Anaphase II** – the separated chromosomes move to opposite ends of the cell.
8. **Telophase II** – cytokinesis occurs, resulting in four haploid daughter cells.

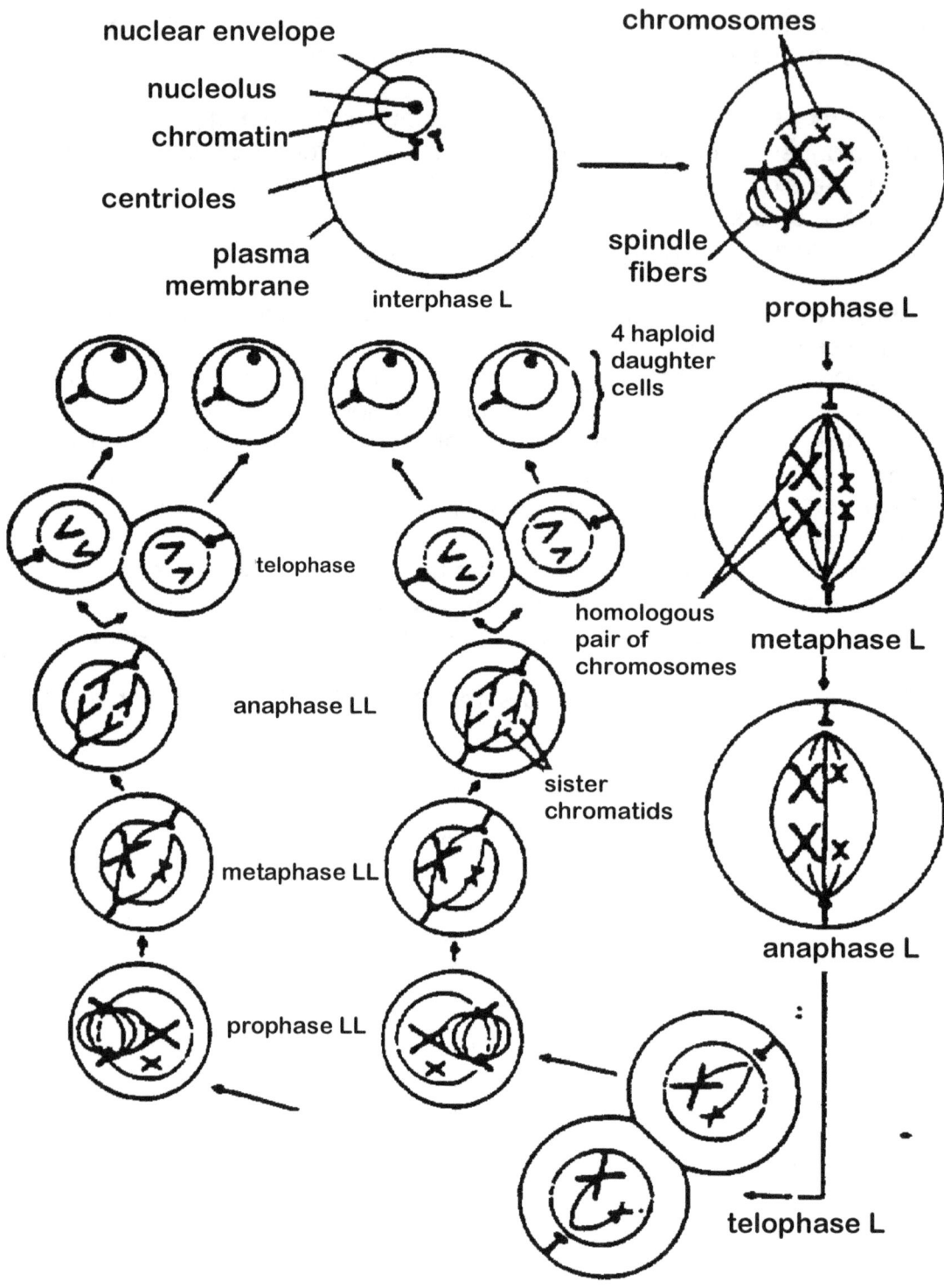

Skill 6.4 **Explain the functions and applications of the instruments and technologies used to study the life sciences at the molecular and cellular levels.**

Gel electrophoresis is a method for analyzing DNA. Electrophoresis separates DNA or protein by size or electrical charge. The DNA runs towards the positive charge as it separates the DNA fragments by size. The gel is treated with a DNA-binding dye that fluoresces under ultraviolet light. A picture of the gel can be taken and used for analysis.

One of the most widely used genetic engineering techniques is **polymerase chain reaction (PCR)**. PCR is a technique in which a piece of DNA can be amplified into billions of copies within a few hours. This process requires primer to specify the segment to be copied, and an enzyme (usually taq polymerase) to amplify the DNA. PCR has allowed scientists to perform several procedures on the smallest amount of DNA.

COMPETENCY 7.0 UNDERSTAND AND APPLY KNOWLEDGE OF THE PRINCIPLES OF HEREDITY AND BIOLOGICAL EVOLUTION.

Skill 7.1 Recognize the nature and function of the gene, with emphasis on the molecular basis of inheritance and gene expression.

Gregor Mendel is recognized as the father of genetics. His work in the late 1800s is the basis of our knowledge of genetics. Although unaware of the presence of DNA or genes, Mendel realized there were factors (now known as **genes**) that were transferred from parents to their offspring. Mendel worked with pea plants and fertilized the plants himself, keeping track of subsequent generations which led to the Mendelian laws of genetics. Mendel found that two "factors" governed each trait, one from each parent. Traits or characteristics came in several forms, known as **alleles**. For example, the trait of flower color had white alleles (*pp*) and purple alleles (*PP*). Mendel formed two laws: the law of segregation and the law of independent assortment.

In bacterial cells, the *lac* operon is a good example of the control of gene expression. The *lac* operon contains the genes that encode for the enzymes used to convert lactose into fuel (glucose and galactose). The *lac* operon contains three genes, *lac Z*, *lac Y*, and *lac A*. *Lac Z* encodes an enzyme for the conversion of lactose into glucose and galactose. *Lac Y* encodes for an enzyme that causes lactose to enter the cell. *Lac A* encodes for an enzyme that acetylates lactose.

The *lac* operon also contains a promoter and an operator that is the "off and on" switch for the operon. A protein called the repressor switches the operon off when it binds to the operator. When lactose is absent, the repressor is active and the operon is turned off. The operon is turned on again when allolactose (formed from lactose) inactivates the repressor by binding to it.

Skill 7.2 Analyze the transmission of genetic information (e.g., Punnett squares, sex-linked traits, pedigree analysis).

The **law of segregation** states that only one of the two possible alleles from each parent is passed on to the offspring. If the two alleles differ, then one is fully expressed in the organism's appearance (the dominant allele) and the other has no noticeable effect on appearance (the recessive allele). The two alleles for each trait segregate into different gametes. A Punnet square can be used to show the law of segregation. In a Punnet square, one parent's genes are put at the top of the box and the other parent's on the side. Genes combine in the squares just like numbers are added in addition tables. This Punnet square shows the result of the cross of two F_1 hybrids.

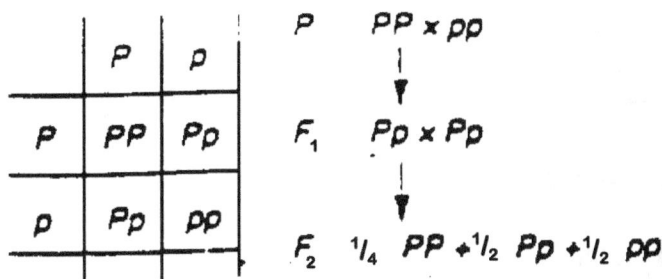

This cross results in a 1:2:1 ratio of F_2 offspring. Here, the P is the dominant allele and the p is the recessive allele. The F_1 cross produces three offspring with the dominant allele expressed (two PP and Pp) and one offspring with the recessive allele expressed (pp). Some other important terms to know:

Homozygous – having a pair of identical alleles. For example, PP and pp are homozygous pairs.
Heterozygous – having two different alleles. For example, Pp is a heterozygous pair.
Phenotype – the organism's physical appearance.
Genotype – the organism's genetic makeup. For example, PP and Pp have the same phenotype (purple in color), but different genotypes.

The **law of independent assortment** states that alleles sort independently of each other. The law of segregation applies for a monohybrid crosses (only one character, in this case flower color, is experimented with). In a dihybrid cross, two characters are being explored. Two of the seven characters Mendel studied were seed shape and color. Yellow is the dominant seed color (Y) and green is the recessive color (y). The dominant seed shape is round (R) and the recessive shape is wrinkled (r). A cross between a plant with yellow round seeds ($YYRR$) and a plant with green wrinkled seeds ($yyrr$) produces an F_1 generation with the genotype $YyRr$. The production of F_2 offspring results in a 9:3:3:1 phenotypic ratio.

	YR	Yr	yR	yr
YR	YYRR	YYRr	YyRR	YyRr
Yr	YYRr	YYrr	YyRr	Yyrr
yR	YyRR	YyRr	yyRR	yyRr
yr	YyRr	Yyrr	yyRr	yyrr

P   YYRR × yyrr

↓

F_1 YyRr

↓

F_2 YYRR – 1
    YYRr – 2
    YyRR – 2 } 9 yellow round
    YyRr – 4

    yyRR – 1
    yyRr – 2 } 3 green round

    YYrr – 1
    Yyrr – 2 } 3 yellow wrinkled

    yyrr – 1 } 1 green wrinkled

Based on Mendelian genetics, the more complex hereditary pattern of **dominance** was discovered. In Mendel's law of segregation, the F_1 generation have either purple or white flowers. This is an example of **complete dominance**. **Incomplete dominance** is when the F_1 generation results in an appearance somewhere between the two parents. For example, red flowers are crossed with white flowers, resulting in an F_1 generation with pink flowers. The red and white traits are still carried by the F_1 generation, resulting in an F_2 generation with a phenotypic ration of 1:2:1. In **codominance,** the genes may form new phenotypes. The ABO blood grouping is an example of codominance. A and B are of equal strength and O is recessive. Therefore, type A blood may have the genotypes of AA or AO, type B blood may have the genotypes of BB or BO, type AB blood has the genotype A and B, and type O blood has two recessive O genes.

PHYSICS

A **family pedigree** is a collection of a family's history for a particular trait. As you work your way through the pedigree of interest, the Mendelian inheritance theories are applied. In tracing a trait, the generations are mapped in a pedigree chart, similar to a family tree but with the alleles present. In a case where both parent have a particular trait and one of two children also express this trait, then the trait is due to a dominant allele. In contrast, if both parents do not express a trait and one of their children do, that trait is due to a recessive allele.

Sex linked traits - the Y chromosome found only in males (XY) carries very little genetic information, whereas the X chromosome found in females (XX) carries very important information. Since men have no second X chromosome to cover up a recessive gene, the recessive trait is expressed more often in men. Women need the recessive gene on both X chromosomes to show the trait. Examples of sex linked traits include hemophilia and color-blindness.

Skill 7.3   Analyze the processes of change at the microscopic and macroscopic levels.

In order to fully understand heredity and biological evolution, students need to comprehend the material on both a smaller (microscopic) and a larger (macroscopic) scale. For example, smaller items would include molecules, DNA, and genes and larger items might include organisms and the biosphere.

The teaching of molecular biology is important to the understanding of the chemical basis of life. Students tend to associate molecules with physical science, not realizing that living systems are made of molecules as well as cells. At the macroscopic level, students learn species as a basis for classifying organisms; in order to understand species at the microscopic level, they need to comprehend the genetic basis of the species.

Microscopic Level:

Atoms, molecules, chemical processes and reactions, bacteria, viruses, protists, cells, tissues, chromosomes, genes, meiosis, mitosis, mutations, comparative embryology, molecular evolution.

Macroscopic Level:

Comparative anatomy, natural selection, convergent evolution, divergent evolution, taxonomy, organs, organisms, body systems, animal and plant structure and function, animal behavior, populations, communities, food chains and webs, biomes.

Skill 7.4 Identify scientific evidence from various sources, such as the fossil record, comparative anatomy, and biochemical similarities, to demonstrate knowledge of theories about processes of biological evolution.

The hypothesis that life developed on Earth from nonliving materials is the most widely accepted theory on the origin of life. The transformation from nonliving materials to life had four stages. The first stage was the nonliving (abiotic) synthesis of small monomers such as amino acids and nucleotides. In the second stage, these monomers combine to form polymers, such as proteins and nucleic acids. The third stage is the accumulation of these polymers into droplets called protobionts. The last stage is the origin of heredity, with RNA as the first genetic material.

The first stage of this theory was hypothesized in the 1920s. A. I. Oparin and J. B. S. Haldane were the first to theorize that the primitive atmosphere was a reducing atmosphere with no oxygen present. The gases were rich in hydrogen, methane, water and ammonia.. In the 1950s, Stanley Miller proved Oparin's theory in the laboratory by combining the above gases. When given an electrical spark, he was able to synthesize simple amino acids. It is commonly accepted that amino acids appeared before DNA. Other laboratory experiments have supported the other stages in the origin of life theory could have happened.

Other scientists believe simpler hereditary systems originated before nucleic acids. In 1991, Julius Rebek was able to synthesize a simple organic molecule that replicates itself. According to his theory, this simple molecule may be the precursor of RNA.

Prokaryotes are the simplest life form. Their small genome size limits the number of genes that control metabolic activities. Over time, some prokaryotic groups became multicellular organisms for this reason. Prokaryotes then evolved to form complex bacterial communities where species benefit from one another.

The **endosymbiotic theory** of the origin of eukaryotes states that eukaryotes arose from symbiotic groups of prokaryotic cells. According to this theory, smaller prokaryotes lived within larger prokaryotic cells, eventually evolving into chloroplasts and mitochondria. Chloroplasts are the descendant of photosynthetic prokaryotes and mitochondria are likely to be the descendants of bacteria that were aerobic heterotrophs. Serial endosymbiosis is a sequence of endosymbiotic events. Serial endosymbiosis may also play a role in the progression of life forms to become eukaryotes.

Fossils are the key to understanding biological history. They are the preserved remnants left by an organism that lived in the past. Scientists have established the geological time scale to determine the age of a fossil. The geological time scale is broken down into four eras: the Precambrian, Paleozoic, Mesozoic, and Cenozoic. The eras are further broken down into periods that represent a distinct age in the history of Earth and its life. Scientists use rock layers called strata to date fossils. The older layers of rock are at the bottom. This allows scientists to correlate the rock layers with the era they date back to. Radiometric dating is a more precise method of dating fossils. Rocks and fossils contain isotopes of elements accumulated over time. The isotope's half-life is used to date older fossils by determining the amount of isotope remaining and comparing it to the half-life.

Dating fossils is helpful to construct and evolutionary tree. Scientists can arrange the succession of animals based on their fossil record. The fossils of an animal's ancestors can be dated and placed on its evolutionary tree. For example, the branched evolution of horses shows the progression of the modern horse's ancestors to be larger, to have a reduced number of toes, and have teeth modified for grazing.

Comparative anatomical studies reveal that some structural features are basically similar – e.g., flowers generally have sepals, petals, stigma, style and ovary but the size, color, number of petals, sepals etc., may differ from species to species.

The degree of resemblance between two organisms indicates how closely they are related in evolution.

- Groups with little in common are supposed to have diverged from a common ancestor much earlier in geological history than groups which have more in common

- To decide how closely two organisms are, anatomists look for the structures which may serve different purpose in the adult, but are basically similar (homologous)

- In cases where similar structures serve different functions in adults, it is important to trace their origin and embryonic development

When a group of organisms share a homologous structure, which is specialized, to perform a variety of functions in order to adapt to different environmental conditions are called adaptive radiation. The gradual spreading of organisms with adaptive radiation is known as divergent evolution.

Examples of divergent evolution are – pentadactyl limb and insect mouthparts

Under similar environmental conditions, fundamentally different structures in different groups of organisms may undergo modifications to serve similar functions. This is called convergent evolution. The structures, which have no close phylogenetic links but showing adaptation to perform the same functions, are called analogous.

Examples are – wings of bats, bird and insects, jointed legs of insects and vertebrates, eyes of vertebrates and cephalopods.

Vestigial organs: Organs that are smaller and simpler in structure than corresponding parts in the ancestral species are called vestigial organs. They are usually degenerated or underdeveloped. These were functional in ancestral species but no have become non functional, e.g., vestigial hind limbs of whales, vestigial leaves of some xerophytes, vestigial wings of flightless birds like ostriches, etc.

TEACHER CERTIFICATION STUDY GUIDE

COMPETENCY 8.0 UNDERSTAND AND APPLY KNOWLEDGE OF THE CHARACTERISTICS AND LIFE FUNCTIONS OF ORGANISMS.

Skill 8.1 Identify the levels of organization of various types of organisms and the structures and functions of cells, tissues, organs, and organ systems.

Life has defining properties. Some of the more important processes and properties associated with life are as follows:

*Order – an organism's complex organization.
*Reproduction – life only comes from life (biogenesis).
*Energy utilization – organisms use and make energy to do many kinds of work.
*Growth and development – DNA directed growth and development.
*Adaptation to the environment – occurs by homeostasis (ability to maintain a certain status), response to stimuli, and evolution.

Life is highly organized. The organization of living systems builds on levels from small to increasingly more large and complex. All aspects, whether it is a cell or an ecosystem, have the same requirements to sustain life. Life is organized from simple to complex in the following way:

Atoms ⑧ molecules ⑧ organelles ⑧ cells ⑧ tissues ⑧ organs ⑧ organ systems ⑧ organism

Skill 8.2 Analyze the strategies and adaptations used by organisms to obtain the basic requirements of life.

Members of the five different kingdoms of the classification system of living organisms often differ in their basic life functions. Here we compare and analyze how members of the five kingdoms obtain nutrients, excrete waste, and reproduce.

Bacteria are prokaryotic, single-celled organisms that lack cell nuclei. The different types of bacteria obtain nutrients in a variety of ways. Most bacteria absorb nutrients from the environment through small channels in their cell walls and membranes (chemotrophs) while some perform photosynthesis (phototrophs). Chemoorganotrophs use organic compounds as energy sources while chemolithotrophs can use inorganic chemicals as energy sources. Depending on the type of metabolism and energy source, bacteria release a variety of waste products (e.g. alcohols, acids, carbon dioxide) to the environment through diffusion.

PHYSICS

All bacteria reproduce through binary fission (asexual reproduction) producing two identical cells. Bacteria reproduce very rapidly, dividing or doubling every twenty minutes in optimal conditions. Asexual reproduction does not allow for genetic variation, but bacteria achieve genetic variety by absorbing DNA from ruptured cells and conjugating or swapping chromosomal or plasmid DNA with other cells.

Animals are multicellular, eukaryotic organisms. All animals obtain nutrients by eating food (ingestion). Different types of animals derive nutrients from eating plants, other animals, or both. Animal cells perform respiration that converts food molecules, mainly carbohydrates and fats, into energy. The excretory systems of animals, like animals themselves, vary in complexity. Simple invertebrates eliminate waste through a single tube, while complex vertebrates have a specialized system of organs that process and excrete waste.

Most animals, unlike bacteria, exist in two distinct sexes. Members of the female sex give birth or lay eggs. Some less developed animals can reproduce asexually. For example, flatworms can divide in two and some unfertilized insect eggs can develop into viable organisms. Most animals reproduce sexually through various mechanisms. For example, aquatic animals reproduce by external fertilization of eggs, while mammals reproduce by internal fertilization. More developed animals possess specialized reproductive systems and cycles that facilitate reproduction and promote genetic variation.

Plants, like animals, are multi-cellular, eukaryotic organisms. Plants obtain nutrients from the soil through their root systems and convert sunlight into energy through photosynthesis. Many plants store waste products in vacuoles or organs (e.g. leaves, bark) that are discarded. Some plants also excrete waste through their roots.

More than half of the plant species reproduce by producing seeds from which new plants grow. Depending on the type of plant, flowers or cones produce seeds. Other plants reproduce by spores, tubers, bulbs, buds, and grafts. The flowers of flowering plants contain the reproductive organs. Pollination is the joining of male and female gametes that is often facilitated by movement by wind or animals.

Fungi are eukaryotic, mostly multi-cellular organisms. All fungi are heterotrophs, obtaining nutrients from other organisms. More specifically, most fungi obtain nutrients by digesting and absorbing nutrients from dead organisms. Fungi secrete enzymes outside of their body to digest organic material and then absorb the nutrients through their cell walls.

Most fungi can reproduce asexually and sexually. Different types of fungi reproduce asexually by mitosis, budding, sporification, or fragmentation. Sexual reproduction of fungi is different from sexual reproduction of animals. The two mating types of fungi are plus and minus, not male and female. The fusion of hyphae, the specialized reproductive structure in fungi, between plus and minus types produces and scatters diverse spores.

Protists are eukaryotic, single-celled organisms. Most protists are heterotrophic, obtaining nutrients by ingesting small molecules and cells and digesting them in vacuoles. All protists reproduce asexually by either binary or multiple fission. Like bacteria, protists achieve genetic variation by exchange of DNA through conjugation.

Skill 8.3 Analyze factors (e.g., physiological, behavioral) that influence homeostasis within an organism.

Animal behavior is responsible for courtship leading to mating, communication between species, territoriality, and aggression between animals and dominance within a group. Behaviors may include body posture, mating calls, display of feathers or fur, coloration or bearing of teeth and claws.

Innate behaviors are inborn or instinctual. An environmental stimulus such as the length of day or temperature results in a behavior. Hibernation among some animals is an innate behavior, as is a change in color known as camouflage.

Learned behavior is modified behavior due to past experience.

Skill 8.4 Demonstrate an understanding of the human as a living organism with life functions comparable to those of other life forms.

Humans are living organisms that display the basic properties of life. Humans share functional characteristics with all living organisms, from simple bacteria to complex mammals. The basic functions of living organisms include reproduction, growth and development, metabolism, and homeostasis/response to the environment.

Reproduction – All living organisms reproduce their own kind. Life arises only from other life. Humans reproduce through sexual reproduction, requiring the interaction of a male and a female. Human sexual reproduction is nearly identical to reproduction in other mammals. In addition, while simpler organisms have different methods of reproduction, they all reproduce. For example, the major mechanism of bacterial reproduction is asexual binary fission in which the cell divides in half, producing two identical cells.

Growth and Development – Growth and development, as directed by DNA, produces an organism characteristic of its species. In humans and other higher-level mammals, growth and development is a very complex process. In humans, growth and development requires differentiation of cells into many different types to form the various organs, structures, and functional elements. While differentiation is unique to higher level organisms, all living organisms grow. For example, the simplest bacterial cell grows in size until it divides into two organisms. Human body cells undergo a similar process, growing in size until division is necessary.

Metabolism – Metabolism is the sum of all chemical reactions that occur in a living organism. Catabolism is the breaking down of complex molecules to release energy. Anabolism is the utilization of the energy from catabolism to build complex molecules. Cellular respiration, the basic mechanism of catabolism in humans, is common to many living organisms of varying levels of complexity.

Homeostasis/Response to the Environment – All living organisms respond and adapt to their environments. Homeostasis is the result of regulatory mechanisms that help maintain an organism's internal environment within tolerable limits. For example, in humans and mammals, constriction and dilation of blood vessels near the skin help maintain body temperature.

TEACHER CERTIFICATION STUDY GUIDE

COMPETENCY 9.0 UNDERSTAND AND APPLY KNOWLEDGE OF HOW ORGANISMS INTERACT WITH EACH OTHER AND WITH THEIR ENVIRONMENT.

Skill 9.1 Identify living and nonliving components of the environment and how they interact with one another.

Succession is an orderly process of replacing a community that has been damaged or has begun where no life previously existed. Primary succession occurs where life never existed before, as in a flooded area or a new volcanic island. Secondary succession takes place in communities that were once flourishing but disturbed by some source, either man or nature, but not totally stripped. A climax community is a community that is established and flourishing.

Abiotic and biotic factors play a role in succession. **Biotic factors** are living things in an ecosystem: plants, animals, bacteria, fungi, etc. **Abiotic factors** are non-living aspects of an ecosystem: soil quality, rainfall, temperature, etc.

Abiotic factors affect succession by way of the species that colonize the area. Certain species will or will not survive depending on the weather, climate, or soil makeup. Biotic factors such as inhibition of one species due to another may occur. This may be due to some form of competition between the species.

Skill 9.2 Recognize the concepts of populations, communities, ecosystems, and ecoregions and the role of biodiversity in living systems.

A **population** is a group of individuals of one species that live in the same general area. Many factors can affect the population size and its growth rate. Population size can depend on the total amount of life a habitat can support. This is the carrying capacity of the environment. Once the habitat runs out of food, water, shelter, or space, the carrying capacity decreases, and then stabilizes.

Skill 9.3 Analyze factors (e.g., ecological, behavioral) that influence interrelationships among organisms.

Ecological and behavioral factors affect the interrelationships among organisms in many ways. Two important ecological factors are environmental conditions and resource availability. Important types of organismal behaviors include competitive, instinctive, territorial, and mating.

Environmental conditions, such as climate, influence organismal interrelationships by changing the dynamic of the ecosystem. Changes in climate such as moisture levels and temperature can alter the environment, changing the characteristics that are advantageous. For example, an increase in temperature will favor those organisms that can tolerate the temperature change. Thus, those organisms gain a competitive advantage. In addition, the availability of necessary resources influences interrelationships. For example, when necessary resources are scarce, interrelationships are more competitive than when resources are abundant.

Types of behavior that influence interrelationships:

Competitive – As previously mentioned, organisms compete for scarce resources. In addition, organisms compete with members of their own species for mates and territory. Many competitive behaviors involve rituals and dominance hierarchies. Rituals are symbolic activities that often settle disputes without undue harm. For example, dogs bare their teeth, erect their ears, and growl to intimidate competitors. A dominance hierarchy, or "pecking order", organizes groups of animals, simplifying interrelationships, conserving energy, and minimizing the potential for harm in a community.

Instinctive – Instinctive, or innate, behavior is common to all members of a given species and is genetically preprogrammed. Environmental differences do not affect instinctive behaviors. For example, baby birds of many types and species beg for food by raising their heads and opening their beaks.

Territorial – Many animals act aggressively to protect their territory from other animals. Animals protect territories for use in feeding, mating, and rearing of young.

Mating – Mating behaviors are very important interspecies interactions. The search for a mate with which to reproduce is an instinctive behavior. Mating interrelationships often involve ritualistic and territorial behaviors that are often competitive.

Skill 9.4 Develop a model or explanation that shows the relationships among organisms in the environment (e.g., food web, food chain, ecological pyramid).

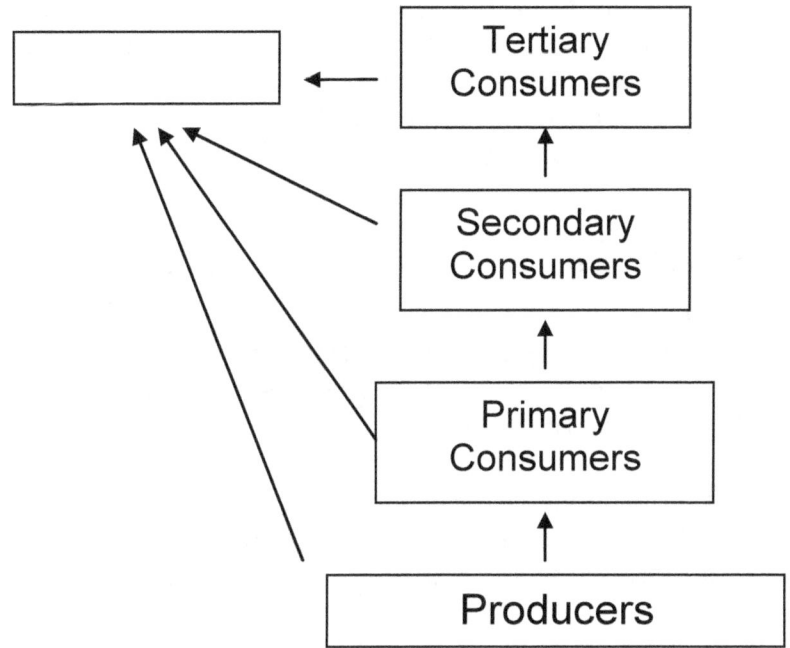

Skill 9.5 Recognize the dynamic nature of the environment, including how communities, ecosystems, and ecoregions change over time.

The environment is ever changing because of natural events and the actions of humans, animals, plants, and other organisms. Even the slightest changes in environmental conditions can greatly influence the function and balance of communities, ecosystems, and ecoregions. For example, subtle changes in salinity and temperature of ocean waters over time can greatly influence the range and population of certain species of fish. In addition, a slight increase in average atmospheric temperature can promote tree growth in a forest, but a corresponding increase in the viability of pathogenic bacteria can decrease the overall growth and productivity of the forest.

Another important concept in ecological change is succession. Ecological succession is the transition in the composition of species in an ecosystem, often after an ecological disturbance in the community. Primary succession begins in an environment virtually void of life, such as a volcanic island. Secondary succession occurs when a natural event disrupts an ecosystem, leaving the soil intact. An example of secondary succession is the reestablishment of a forest after destruction by a forest fire.

PHYSICS 65

Factors that drive the process of succession include interspecies competition, environmental conditions, inhibition, and facilitation. In a developing ecosystem, species compete for scarce resources. The species that compete most successfully dominate. Environmental conditions, as previously discussed, influence species viability. Finally, the activities of certain species can inhibit or facilitate the growth and development of other species. Inhibition results from exploitative competition or interference competition. In facilitation, a species or group of species lays the foundation for the establishment of other, more advanced species. For example, the presence of a certain bacterial population can change the pH of the soil, allowing for the growth of different types of plants and trees.

Skill 9.6 Analyze interactions of humans with their environment.

The human population has been growing exponentially for centuries. People are living longer and healthier lives than ever before. Better health care and nutrition practices have helped in the survival of the population.

Human activity affects parts of the nutrient cycles by removing nutrients from one part of the biosphere and adding them to another. This results in nutrient depletion in one area and nutrient excess in another. This affects water systems, crops, wildlife, and humans.

Humans are responsible for the depletion of the ozone layer. This depletion is due to chemicals used for refrigeration and aerosols. The consequences of ozone depletion will be severe. Ozone protects the Earth from the majority of UV radiation. An increase of UV will promote skin cancer and unknown effects on wildlife and plants.

Humans have a tremendous impact on the world's natural resources. The world's natural water supplies are affected by human use. Waterways are major sources for recreation and freight transportation. Oil and wastes from boats and cargo ships pollute the aquatic environment. The aquatic plant and animal life is affected by this contamination.

Deforestation for urban development has resulted in the extinction or relocation of several species of plants and animals. Animals are forced to leave their forest homes or perish amongst the destruction. The number of plant and animal species that have become extinct due to deforestation is unknown. Scientists have only identified a fraction of the species on Earth. It is known that if the destruction of natural resources continues, there may be no plants or animals successfully reproducing in the wild.

Humans are continuously searching for new places to form communities. This encroachment on the environment leads to the destruction of wildlife communities. Conservationists focus on endangered species, but the primary focus should be on protecting the entire biome. If a biome becomes extinct, the wildlife dies or invades another biome.

Preservations established by the government aim at protecting small parts of biomes. While beneficial in the conservation of a few areas, the majority of the environment is still unprotected.

Skill 9.7 Explain the functions and applications of the instruments and technologies used to study the life sciences at the organism and ecosystem level.

Biologists use a variety of tools and technologies to perform tests, collect and display data, and analyze relationships at the organismal and ecosystem level. Examples of commonly used tools include computer-linked probes, computerized tracking devices, computer models and databases, and spreadsheets.

Biologists use computer-linked probes to measure various environmental factors including temperature, dissolved oxygen, pH, ionic concentration, and pressure. The advantage of computer-linked probes, as compared to more traditional observational tools, is that the probes automatically gather data and present it in an accessible format. This property of computer-linked probes eliminates the need for constant human observation and manipulation.

Biologists use computerized tracking devices to study the behavior of animals in an ecosystem. Biologists can implant computer chips on animals to track movement and migration, population changes, and general behavioral characteristics.

Computer models allow biologists to use data and information they collect in the field to make predictions and projections about the future of ecosystems and organisms. Because ecosystems are large and change very slowly, direct observation is not a suitable strategy for ecological studies. For example, while a scientist cannot reasonably expect to observe and gather data over an entire ecosystem, she can collect samples and use computer databases and models to make projections about the ecosystem as a whole.

Finally, biologists use spreadsheets to organize, analyze, and display data. For example, conservation ecologists use spreadsheets to model population growth and development, apply sampling techniques, and create statistical distributions to analyze relationships. Spreadsheet use simplifies data collection and manipulation and allows the presentation of data in a logical and understandable format.

TEACHER CERTIFICATION STUDY GUIDE

SUBAREA III. **PHYSICAL SCIENCE**

COMPETENCY 10.0 UNDERSTAND AND APPLY KNOWLEDGE OF THE NATURE AND PROPERTIES OF ENERGY IN ITS VARIOUS FORMS.

"Energy is an abstract concept invented by physical scientists in the nineteenth century to describe quantitatively a wide variety of natural phenomena."
                                                                David Rose, MIT

Skill 10.1   **Describe the characteristics of and relationships among thermal, acoustical, radiant, electrical, chemical, mechanical, and nuclear energies through conceptual questions.**

Abstract concept it might be, but energy is one of the most fundamental concepts in our world. We use it to move people and things from place to place, to heat and light our homes, to entertain us, to produce food and goods and to communicate with each other. It is not some sort of magical invisible fluid, poured, weighed or bottled. It is not a substance but rather the ability possessed by things.

Technically, **energy is the ability to do work or supply heat.** Work is the transfer of energy to move an object a certain distance. It is the motion against an opposing force. Lifting a chair into the air is work; the opposing force is gravity. Pushing a chair across the floor is work; the opposing force is friction.

Heat, on the other hand, is not a form of energy but a method of transferring energy.

This energy, according to the First Law of Thermodynamics, is conserved. That means energy is neither created nor destroyed in ordinary physical and chemical processes (non-nuclear). Energy is merely changed from one form to another. Energy in all of its forms must be conserved. In any system, $\Delta E = q + w$ (E = energy, q = heat and w = work).

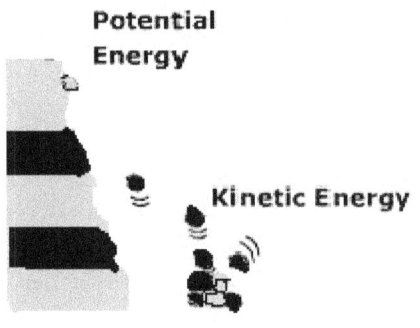

Energy exists in two basic forms, potential and kinetic. Kinetic energy is the energy of a moving object. Potential energy is the energy stored in matter due to position relative to other objects.

In any object, solid, liquid or gas, the atoms and molecules that make up the object are constantly moving (vibrational, translation and rotational motion) and colliding with each other. They are not stationary.

Due to this motion, the object's particles have varying amounts of kinetic energy. A fast moving atom can push a slower moving atom during a collision, so it has energy. All moving objects have energy and that energy depends on the object's mass and velocity. Kinetic energy is calculated: $K.E. = \frac{1}{2} mv^2$.

The temperature exhibited by an object is proportional to the average kinetic energy of the particles in the substance. Increase the temperature of a substance and its particles move faster so their average kinetic energies increase as well. But temperature is NOT an energy, it is not conserved.

The energy an object has due to its position or arrangement of its parts is called potential energy. Potential energy due to position is equal to the mass of the object times the gravitational pull on the object times the height of the object, or:

$$PE = mgh$$

Where PE = potential energy; m = mass of object; g = gravity; and h = height.

Heat is energy that is transferred between objects caused by differences in their temperatures. Heat passes spontaneously from an object of higher temperature to one of lower temperature. This transfer continues until both objects reach the same temperature. Both kinetic energy and potential energy can be transformed into heat energy. When you step on the brakes in your car, the kinetic energy of the car is changed to heat energy by friction between the brake and the wheels. Other transformations can occur from kinetic to potential as well. Since most of the energy in our world is in a form that is not easily used, man and mother nature has developed some clever ways of changing one form of energy into another form that may be more useful.

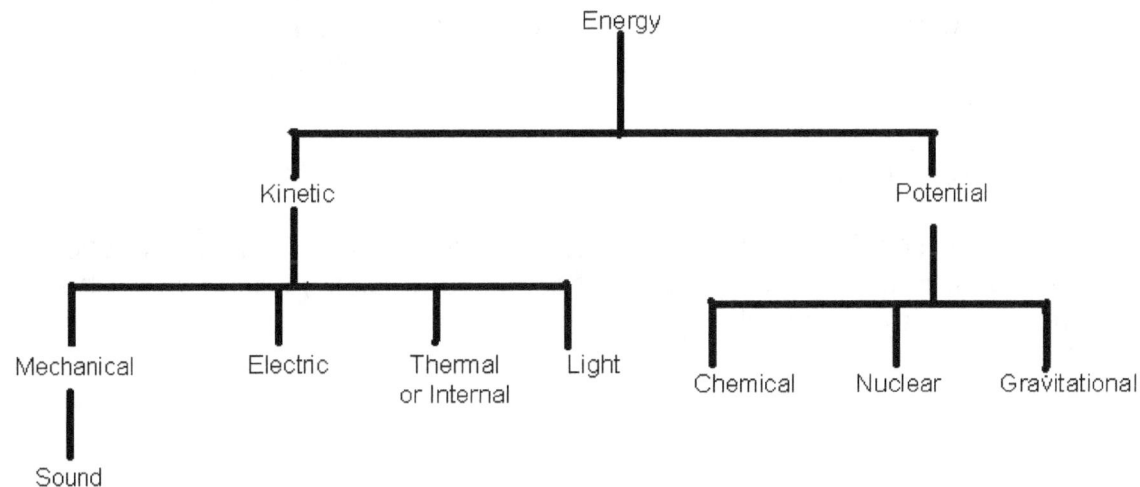

Gravitational Potential Energy:

When something is lifted or suspended in air, work is done on the object against the pull of gravity. This work is converted to a form of potential energy called gravitational potential energy.

Nuclear Potential Energy:

The nuclear energy trapped inside the atom is referred to as nuclear energy. When the atom is split, tremendous energy is released in the form of heat and light.

Chemical Potential Energy:

The energy generated from chemical reactions in which the chemical bonds of a substance are broken and rearranged to form new substances is called chemical potential energy.

Electrical Kinetic energy:

The flow of electrons along a circuit is called electrical energy. The movement of electrons creates an electric current which generates electricity

Mechanical Kinetic Energy:

Mechanical energy is the energy of motion doing work, like a pendulum moving back and forth in a grandfather clock.

Thermal Kinetic Energy:

Thermal Energy is defined as the energy that a substance has due to the chaotic motion of its molecules. Molecules are in constant motion, and always possess some amount of kinetic energy. It is also called Internal energy and is not the same as heat.

Light or Radiant Kinetic Energy:

Radiant energy comes from a light source, such as the sun. Energy released from the sun is in the form of photons. These tiny particles, invisible to the human eye, move in a way similar to a wave.

Energy transformations make it possible for us to use energy; do work. Here are some examples of how energy is transformed to do work:

1. Different types of stoves are used to transform the chemical energy of the fuel (gas, coal, wood, etc.) into heat.

2. Solar collectors can be used to transform solar energy into electrical energy.

3. Wind mills make use of the kinetic energy of the air molecules, transforming it into mechanical or electrical energy.

4. Hydroelectric plants transform the kinetic energy of falling water into electrical energy.

5. A flashlight converts chemical energy stored in batteries to light energy and heat. Most of the energy is converted to heat, only a small amount is actually changed into light energy.

Skill 10.2 Analyze the processes by which energy is exchanged or transformed through conceptual questions.

All energy transformations can be traced back to the sun -- the original source of energy for life on earth. The sun produces heat, light, and radiation through the process of fusion. The sun converts hydrogen to helium in a three-step process.

1. Two hydrogen atoms combine. This forms deuterium, also called heavy hydrogen.
2. Deuterium joins with another hydrogen atom. This forms a type of helium (helium-3).
3. Two helium-3 atoms collide, producing ordinary helium and two hydrogen atoms

Energy is released at every step. The energy comes in the form of tremendous heat, radiation that can kill us, and light that we need to survive. This energy forms in the core of the sun moves through several other layers to the photosphere and is emitted as light.

Once on earth, radiant energy (light) is transformed to the mechanical energy of wind and waves found in nature, through photosynthesis to the stored energy in plants, accumulated as chemical energy in deposits of coal, oil and gas.

Man uses mechanical, thermal and radiant energy to provide for his needs. Mechanical energy operates machines, thermal energy cooks food and heats homes while radiant energy provides light.

Plants capture the sunlight and through photosynthesis change it into chemical energy stored in carbohydrate (sugar) molecules in plant cells. The chlorophyll molecule found in plants captures light energy and uses it to build these carbohydrate molecules from the raw materials water, carbon dioxide and some minerals.

$$CO_2 (g) + H_2O (g) + chlorophyll + sunlight \rightarrow Oxygen (g) + sugar.$$

This process produces oxygen and glucose, $C_6H_{12}O_6$. The sugar (glucose) is stored in the plant.

A person eats the plant. The chemical energy stored in the plant is transferred to the person. Body processes like digestion, circulation and respiration are fueled by cells converting stored chemical energy into work and heat.

Inside muscle cells, the chemical energy is converted to mechanical energy and heat. The muscle twitches and the body jumps to life. Stored chemical energy has been converted into kinetic energy. Energy not used immediately by the cells is stored as potential energy in fat cells.

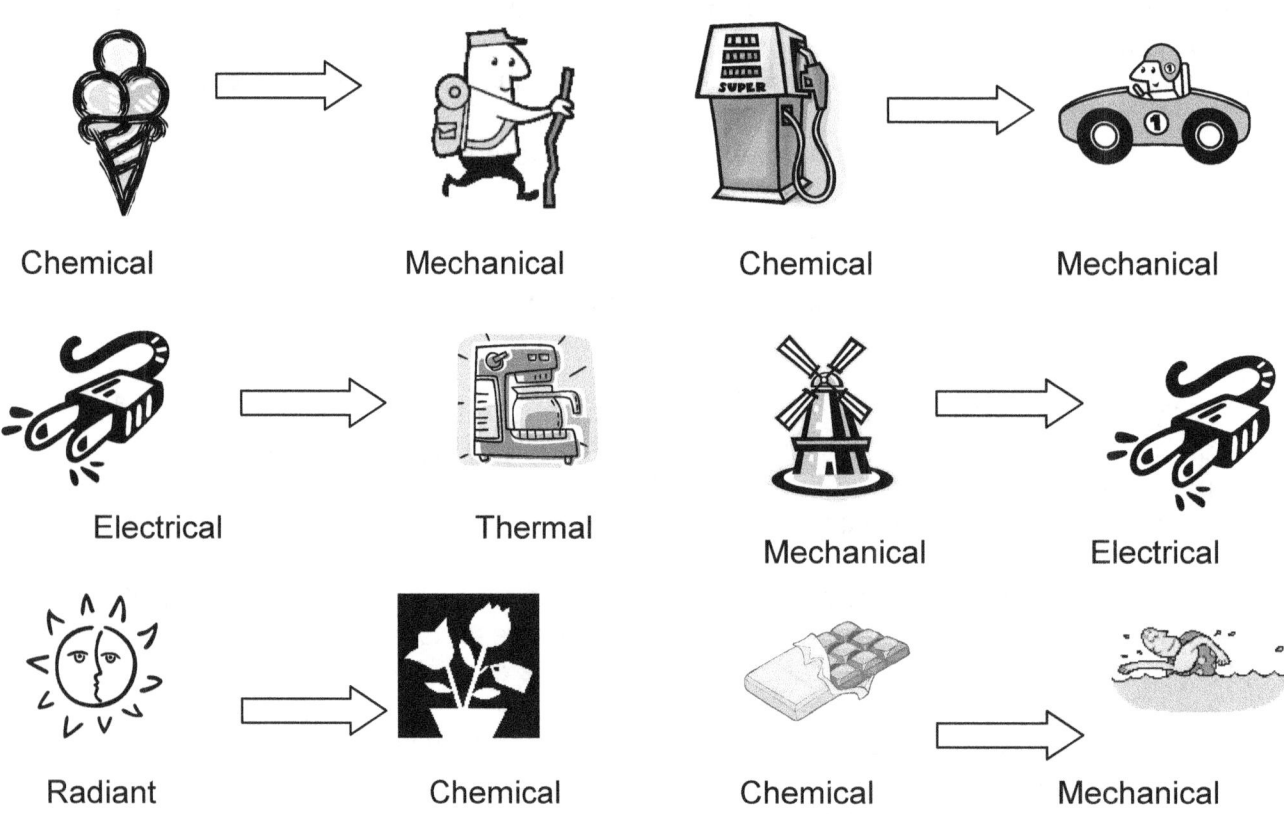

TEACHER CERTIFICATION STUDY GUIDE

Skill 10.3   Apply the three laws of thermodynamics to explain energy transformations, including basic algebraic problem solving

The three laws of thermodynamics are as follows:

1. The total amount of energy in the universe is constant, energy cannot be created or destroyed, but can merely change form.

    Equation:
    $\Delta E = Q + W$
    Change in energy = (Heat energy entering or leaving) + (work done)

2. In energy transformations, entropy (disorder) increases and useful energy is lost (as heat).

    Equation:
    $\Delta S = \Delta Q/T$
    Change in entropy = (Heat transfer) / (Temperature)

3. As the temperature of a system approaches absolute zero, entropy (disorder) approaches a constant.

Sample Problems:

1. A car engine burns gasoline to power the car. An amount of gasoline containing 2000J of stored chemical energy produced 1500J of mechanical energy to power the engine. How much heat energy did the engine release?

Solution:
$\Delta E = Q + W$	the first law of thermodynamics
00J = Q + 1500J	apply the first law
Q (work) = 500J	

2. 18200J of heat leaks out of a hot oven. The temperature of the room is 25°C (298K). What is the increase in entropy resulting from this heat transfer?

Solution:
$\Delta S = \Delta Q/T$	the second law of thermodynamics
ΔS = 18200J / 298K	apply the second law
= 61.1 J/K	solve

PHYSICS

Skill 10.4 Apply the principle of conservation as it applies to energy through conceptual questions and solving basic algebraic problems.

The law of conservation of energy states that energy is neither created nor destroyed. Thus, energy changes form when energy transactions occur in nature. Because the total energy in the universe is constant, energy continually transitions between forms. For example, an engine burns gasoline converting the chemical energy of the gasoline into mechanical energy, a plant converts radiant energy of the sun into chemical energy found in glucose, or a battery converts chemical energy into electrical energy.

TEACHER CERTIFICATION STUDY GUIDE

COMPETENCY 11.0 UNDERSTAND AND APPLY KNOWLEDGE OF THE STRUCTURE AND PROPERTIES OF MATTER.

Skill 11.1 Describe the nuclear and atomic structure of matter, including the three basic parts of the atom.

Experimental evidence shows that the atom is mostly empty space. This empty space is called the electron cloud and it gives the atom its size. In the middle of the electron cloud is the nucleus that contains protons and neutrons. The nucleus gives the atom most of its mass. A proton has a mass of 1 atomic mass unit (amu) and a charge of +1 while a neutron has a similar mass, 1 amu, but no charge. An electron has a very small mass but compared to a proton or neutron it is negligible so it is considered to have no mass. It does, however, have a -1 charge.

Atoms of the same element always have the same number of protons, called the atomic number. However, elements can have isotopes. Isotopes are atoms of the same element but have differing numbers of neutrons have hence, differing atomic masses.

Isotopes are indicated using this symbol:

^1_1H or ^2_1H or ^3_1H for example.

Skill 11.2 Analyze the properties of materials in relation to their chemical or physical structures (e.g., periodic table trends, relationships, and properties) and evaluate uses of the materials based on their properties.

The **periodic table of elements** is an arrangement of the elements in rows and columns so that it is easy to locate elements with similar properties. The elements of the modern periodic table are arranged in numerical order by atomic number.

The **periods** are the rows down the left side of the table. They are called first period, second period, etc. The columns of the periodic table are called **groups**, or **families.** Elements in a family have similar properties.

There are three types of elements that are grouped by color: metals, nonmetals, and metalloids.

PHYSICS

Element Key
Atomic
Number

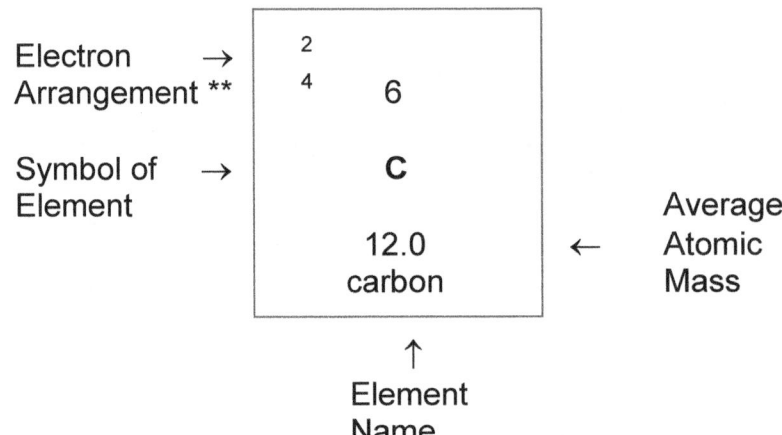

** Number of electrons on each level. Top number represents the innermost level.

The periodic table arranges metals into families with similar properties. The periodic table has its columns marked IA - VIIIA. These are the traditional group numbers. Arabic numbers 1 - 18 are also used, as suggested by the Union of Physicists and Chemists. The Arabic numerals will be used in this text.

Metals:

With the exception of hydrogen, all elements in Group 1 are **alkali metals**. These metals are shiny, softer, and less dense, and the most chemically active.

Group 2 metals are the **alkaline earth metals.** They are harder, denser, have higher melting points, and are chemically active.

The **transition elements** can be found by finding the periods (rows) from 4 to 7 under the groups (columns) 3 - 12. They are metals that do not show a range of properties as you move across the chart. They are hard and have high melting points. Compounds of these elements are colorful, such as silver, gold, and mercury.

Elements can be combined to make metallic objects. An **alloy** is a mixture of two or more elements having properties of metals. The elements do not have to be all metals. For instance, steel is made up of the metal iron and the non-metal carbon.

Nonmetals:

Nonmetals are not as easy to recognize as metals because they do not always share physical properties. However, in general the properties of nonmetals are the opposite of metals. They are not shiny, are brittle, and are not good conductors of heat and electricity.

Nonmetals are solids, gases, and one liquid (bromine).

Nonmetals have four to eight electrons in their outermost energy levels and tend to attract electrons to their outer energy levels. As a result, the outer levels usually are filled with eight electrons. This difference in the number of electrons is what caused the differences between metals and nonmetals. The outstanding chemical property of nonmetals is that react with metals.

The **halogens** can be found in Group 17. Halogens combine readily with metals to form salts. Table salt, fluoride toothpaste, and bleach all have an element from the halogen family.

The **Noble Gases** got their name from the fact that they did not react chemically with other elements, much like the nobility did not mix with the masses. These gases (found in Group 18) will only combine with other elements under very specific conditions. They are **inert** (inactive).

In recent years, scientists have found this to be only generally true, since chemists have been able to prepare compounds of krypton and xenon.

Metalloids:

Metalloids have properties in between metals and nonmetals. They can be found in Groups 13 - 16, but do not occupy the entire group. They are arranged in stair steps across the groups.

Physical Properties:
1. All are solids having the appearance of metals.
2. All are white or gray, but not shiny.
3. They will conduct electricity, but not as well as a metal.

Chemical Properties:
1. Have some characteristics of metals and nonmetals.
2. Properties do not follow patterns like metals and nonmetals. Each must be studied individually.

Boron is the first element in Group 13. It is a poor conductor of electricity at low temperatures. However, increase its temperature and it becomes a good conductor. By comparison, metals, which are good conductors, lose their ability as they are heated. It is because of this property that boron is so useful. Boron is a semiconductor. **Semiconductors** are used in electrical devices that have to function at temperatures too high for metals.

Silicon is the second element in Group 14. It is also a semiconductor and is found in great abundance in the earth's crust. Sand is made of a silicon compound, silicon dioxide. Silicon is also used in the manufacture of glass and cement.

Skill 11.3 **Apply the principle of conservation as it applies to mass and charge through conceptual questions.**

The principle of conservation states that certain measurable properties of an isolated system remain constant despite changes in the system. Two important principles of conservation are the conservation of mass and charge.

The principle of conservation of mass states that the total mass of a system is constant. Examples of conservation in mass in nature include the burning of wood, rusting of iron, and phase changes of matter. When wood burns, the total mass of the products, such as soot, ash, and gases, equals the mass of the wood and the oxygen that reacts with it. When iron reacts with oxygen, rust forms. The total mass of the iron-rust complex does not change. Finally, when matter changes phase, mass remains constant. Thus, when a glacier melts due to atmospheric warming, the mass of liquid water formed is equal to the mass of the glacier.

The principle of conservation of charge states that the total electrical charge of a closed system is constant. Thus, in chemical reactions and interactions of charged objects, the total charge does not change. Chemical reactions and the interaction of charged molecules are essential and common processes in living organisms and systems.

Skill 11.4   Analyze bonding and chemical, atomic, and nuclear reactions (including endothermic and exothermic reactions) in natural and man-made systems and apply basic stoichiometric principles.

Chemical reactions are the interactions of substances resulting in chemical change and change in energy. Chemical reactions involve changes in electron motion and the breaking and forming of chemical bonds. Reactants are the original substances that interact to form distinct products. Endothermic chemical reactions consume energy while exothermic chemical reactions release energy with product formation. Chemical reactions occur continually in nature and are also induced by man for many purposes.

Nuclear reactions, or **atomic reactions**, are reactions that change the composition, energy, or structure of atomic nuclei. Nuclear reactions change the number of protons and neutrons in the nucleus. The two main types of nuclear reactions are fission (splitting of nuclei) and fusion (joining of nuclei). Fusion reactions are exothermic, releasing heat energy. Fission reactions are endothermic, absorbing heat energy. Fission of large nuclei (e.g. uranium) releases energy because the products of fission undergo further fusion reactions. Fission and fusion reactions can occur naturally, but are most recognized as man-made events. Particle acceleration and bombardment with neutrons are two methods of inducing nuclear reactions.

Stoichiometry is the calculation of quantitative relationships between reactants and products in chemical reactions. Scientists use stoichiometry to balance chemical equations, make conversions between units of measurement (e.g. grams to moles), and determine the correct amount of reactants to use in chemical reactions.

Example:

The reaction of iron (Fe) and hydrochloric acid (HCl) produces H_2 and $FeCl_2$. Determine the amount of HCl required to react with 200g of Fe.

$Fe + HCl = H_2 + FeCl_2$

$Fe + 2HCl = H_2 + FeCl_2$ Balance equation (equal number of atoms on each side)

$$\frac{200 g\ Fe}{1} \frac{1\ mol\ Fe}{55.8 g\ Fe} \frac{2\ mol\ HCl}{1\ mol\ Fe} \frac{36.5 g\ HCl}{1\ mol\ HCl}$$ Perform stoichiometric calculations

= 262 g of HCl required to react completely with 200 g Fe   Solve

Skill 11.5 Apply kinetic theory to explain interactions of energy with matter, including conceptual questions on changes in state.

The kinetic theory states that matter consists of molecules, possessing kinetic energies, in continual random motion. The state of matter (solid, liquid, or gas) depends on the speed of the molecules and the amount of kinetic energy the molecules possess. The molecules of solid matter merely vibrate allowing strong intermolecular forces to hold the molecules in place. The molecules of liquid matter move freely and quickly throughout the body and the molecules of gaseous matter move randomly and at high speeds.

Matter changes state when energy is added or taken away. The addition of energy, usually in the form of heat, increases the speed and kinetic energy of the component molecules. Faster moving molecules more readily overcome the intermolecular attractions that maintain the form of solids and liquids. In conclusion, as the speed of molecules increases, matter changes state from solid to liquid to gas (melting and evaporation).

As matter loses heat energy to the environment, the speed of the component molecules decrease. Intermolecular forces have greater impact on slower moving molecules. Thus, as the speed of molecules decrease, matter changes from gas to liquid to solid (condensation and freezing).

Skill 11.6 Explain the functions and applications of the instruments and technologies used to study matter and energy.

Scientists utilize various instruments and technologies to study matter and energy. Commonly used instruments include spectrometers, basic measuring devices, thermometers, and calorimeters.

Spectroscopy is the study of absorption and emission of energy of different frequencies by molecules and atoms. The spectrometer is the instrument used in spectroscopy. Because different molecules and atoms have different spectroscopic properties, spectroscopy helps scientists determine the molecular composition of matter.

Basic devices, like scales and rulers, measure the physical properties of matter. Scientists often measure the size, volume, and mass of different forms of matter.

Thermometers and calorimeters measure the energy exchanged in chemical reactions. Temperature change during chemical reactions is indicative of the flow of energy into or out of a system of reactants and products.

COMPETENCY 12.0 UNDERSTAND AND APPLY KNOWLEDGE OF FORCES AND MOTION.

Skill 12.1 Demonstrate an understanding of the concepts and interrelationships of position, time, velocity, and acceleration through conceptual questions, algebra-based kinematics, and graphical analysis.

The science of describing the motion of bodies is known as **kinematics**. The motion of bodies is described using words, diagrams, numbers, graphs, and equations.

The following words are used to describe motion: vectors, scalars, distance, displacement, speed, velocity, and acceleration.

The two categories of mathematical quantities that are used to describe the motion of objects are scalars and vectors. **Scalars** are quantities that are fully described by magnitude alone. Examples of scalars are 5m and 20 degrees Celsius. **Vectors** are quantities that are fully described by magnitude and direction. Examples of vectors are 30m/sec, and 5 miles north.

Distance is a scalar quantity that refers to how much ground an object has covered while moving. **Displacement** is a vector quantity that refers to the object's change in position.

Example:

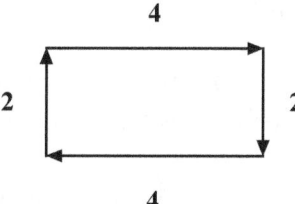

Jamie walked 2 miles north, 4 miles east, 2 miles south, and then 4 miles west. In terms of distance, she walked 12 miles. However, there is no displacement because the directions cancelled each other out, and she returned to her starting position.

Speed is a scalar quantity that refers to how fast an object is moving (ex. the car was traveling 60 mi./hr). **Velocity** is a vector quantity that refers to the rate at which an object changes its position. In other words, velocity is speed with direction (ex. the car was traveling 60 mi./hr east).

$$\text{Average speed} = \frac{\text{Distance traveled}}{\text{Time of travel}}$$

$$v = \frac{d}{t}$$

$$\text{Average velocity} = \frac{\Delta \text{position}}{\text{time}} = \frac{\text{displacement}}{\text{time}}$$

Instantaneous Speed - speed at any given instant in time.

Average Speed - average of all instantaneous speeds, found simply by a distance/time ratio.

Acceleration is a vector quantity defined as the rate at which an object changes its velocity.

$$a = \frac{\Delta velocity}{time} = \frac{v_f - v_i}{t}$$ where *f* represents the final velocity and *i* represents the initial velocity

Since acceleration is a vector quantity, it always has a direction associated with it. The direction of the acceleration vector depends on

- whether the object is speeding up or slowing down
- whether the object is moving in the positive or negative direction.

Skill 12.2 **Demonstrate an understanding of the concepts and interrelationships of force (including gravity and friction), inertia, work, power, energy, and momentum.**

Dynamics is the study of the relationship between motion and the forces affecting motion. **Force** causes motion. Mass and weight are not the same quantities. An object's **mass** gives it a reluctance to change its current state of motion. It is also the measure of an object's resistance to acceleration. The force that the earth's gravity exerts on an object with a specific mass is called the object's weight on earth. Weight is a force that is measured in Newtons. Weight (W) = mass times acceleration due to gravity (**W = mg**). To illustrate the difference between mass and weight, picture two rocks of equal mass on a balance scale. If the scale is balanced in one place, it will be balanced everywhere, regardless of the gravitational field. However, the weight of the stones would vary on a spring scale, depending upon the gravitational field. In other words, the stones would be balanced both on earth and on the moon. However, the weight of the stones would be greater on earth than on the moon.

Surfaces that touch each other have a certain resistance to motion. This resistance is **friction.**
1. The materials that make up the surfaces will determine the magnitude of the frictional force.
2. The frictional force is independent of the area of contact between the two surfaces.
3. The direction of the frictional force is opposite to the direction of motion.
4. The frictional force is proportional to the normal force between the two surfaces in contact.

Static friction describes the force of friction of two surfaces that are in contact but do not have any motion relative to each other, such as a block sitting on an inclined plane. **Kinetic friction** describes the force of friction of two surfaces in contact with each other when there is relative motion between the surfaces.

When an object moves in a circular path, a force must be directed toward the center of the circle in order to keep the motion going. This constraining force is called **centripetal force**. Gravity is the centripetal force that keeps a satellite circling the earth.

Push and pull –Pushing a volleyball or pulling a bowstring applies muscular force when the muscles expand and contract. Elastic force is when any object returns to its original shape (for example, when a bow is released).

Rubbing – Friction opposes the motion of one surface past another. Friction is common when slowing down a car or sledding down a hill.

Pull of gravity – is a force of attraction between two objects. Gravity questions can be raised not only on earth but also between planets and even black hole discussions.

(a) **Forces on objects at rest** – The formula F= m/a is shorthand for force equals mass over acceleration. An object will not move unless the force is strong enough to move the mass. Also, there can be opposing forces holding the object in place. For instance, a boat may want to be forced by the currents to drift away but an equal and opposite force is a rope holding it to a dock.

Forces on a moving object - Overcoming inertia is the tendency of any object to oppose a change in motion. An object at rest tends to stay at rest. An object that is moving tends to keep moving.

(b) **Inertia and circular motion** – The centripetal force is provided by the high banking of the curved road and by friction between the wheels and the road. This inward force that keeps an object moving in a circle is called centripetal force.

Simple machines include the following:

1. Inclined plane
2. Lever
3. Wheel and axle
4. Pulley

Compound machines are two or more simple machines working together. A wheelbarrow is an example of a complex machine. It uses a lever and a wheel and axle. Machines of all types ease workload by changing the size or direction of an applied force. The amount of effort saved when using simple or complex machines is called mechanical advantage or MA.

Work is done on an object when an applied force moves through a distance.

Power is the work done divided by the amount of time that it took to do it. (Power = Work / time)

Skill 12.3 Describe and predict the motions of bodies in one and two dimensions in inertial and accelerated frames of reference in a physical system, including projectile motion but excluding circular motion.

Newton's Three Laws of Motion:

First Law: An object at rest tends to stay at rest and an object in motion tends to stay in motion with the same speed and in the same direction unless acted upon by an unbalanced force, for example, when riding on a descending elevator that suddenly stops, blood rushes from your head to your feet. **Inertia** is the resistance an object has to a change in its state of motion.

Second Law: The acceleration of an object depends directly upon the net force acting upon the object, and inversely upon the mass of the object. As the net force increases, so will the object's acceleration. However, as the mass of the object increases, its acceleration will decrease.

$$F_{net} = m*a$$

Third Law: For every action, there is an equal and opposite reaction, for example, when a bird is flying, the motion of its wings pushes air downward; the air reacts by pushing the bird upward.

Projectile Motion

By definition, a **projectile** only has one force acting upon it – the force of gravity.

Gravity influences the vertical motion of the projectile, causing vertical acceleration. The horizontal motion of the projectile is the result of the tendency of any object in motion to remain in motion at constant velocity. (Remember, there are no horizontal forces acting upon the projectile. By definition, gravity is the only force acting upon the projectile.)

Projectiles travel with a parabolic trajectory due to the fact that the downward force of gravity accelerates them downward from their otherwise straight-line trajectory. Gravity affects the vertical motion, not the horizontal motion, of the projectile. Gravity causes a downward displacement from the position that the object would be in if there were no gravity.

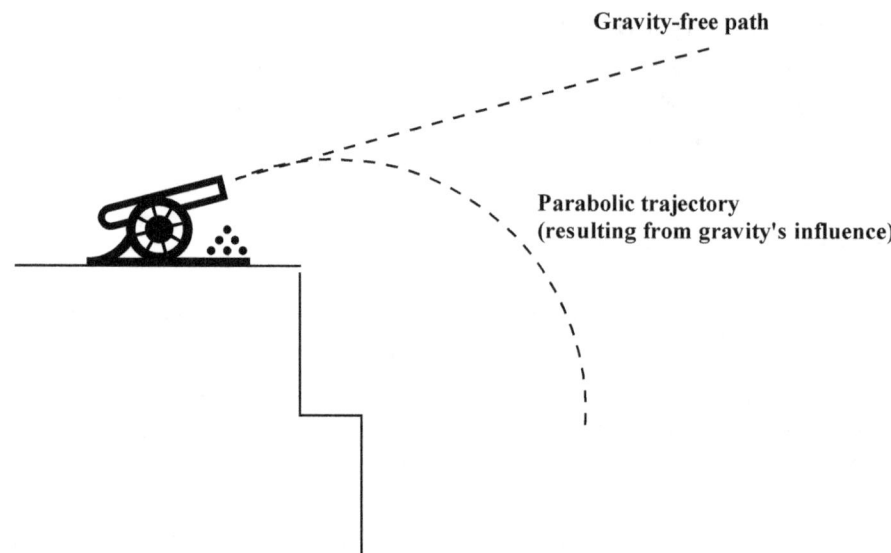

Skill 12.4 Analyze and predict motions and interactions of bodies involving forces within the context of conservation of energy and/or momentum through conceptual questions and algebra-based problem solving.

The Law of **Conservation of Energy** states that energy may neither be created nor destroyed. Therefore, the sum of all energies in the system is a constant.

Example:

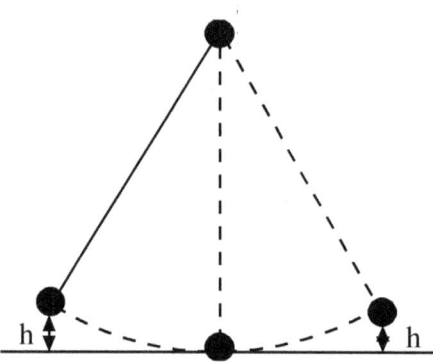

The formula to calculate the potential energy is PE = mgh.

The mass of the ball = 20kg
The height, h = 0.4m
The acceleration due to gravity, g = 9.8 m/s^2

PE = mgh
PE = 20(.4)(9.8)
PE = 78.4J (Joules, units of energy)

The position of the ball on the left is where the Potential Energy (PE) = 78.4J resides while the Kinetic Energy (KE) = 0. As the ball is approaching the center position, the PE is decreasing while the KE is increasing. At exactly halfway between the left and center positions, the PE = KE.

The center position of the ball is where the Kinetic Energy is at its maximum while the Potential Energy (PE) = 0. At this point, theoretically, the entire PE has transformed into KE. Now the KE = 78.4J while the PE = 0.

The right position of the ball is where the Potential Energy (PE) is once again at its maximum and the Kinetic Energy (KE) = 0.

We can now say that:

$$PE + KE = 0$$
$$PE = -KE$$

The sum of PE and KE is the **total mechanical energy:**

Total Mechanical Energy = PE + KE

The law of **momentum conservation** can be stated as follows: For a collision occurring between object 1 and object 2 in an isolated system, the total momentum of the two objects before the collision is equal to the total momentum of the two objects after the collision. That is, the momentum lost by object 1 is equal to the momentum gained by object 2.

Example:

> A 90 kg soccer player moving west at 4 m/s collides with an 80 kg soccer player moving east at 5 m/s. After the collision, both players move east at 3 m/s. Draw a vector diagram in which the before and after collision momenta of each player is represented by a momentum vector.

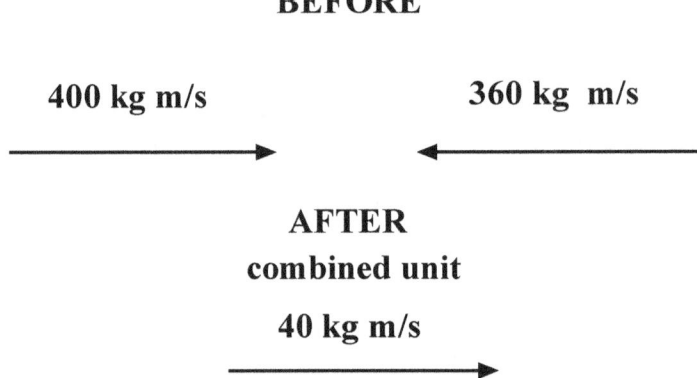

The combined momentum before the collision is +400 kg m/s + (- 360 kg m/s) or 40 kg m/s. According to the law of conservation of momentum, the combined momentum after a collision must be equal to the combined momentum before the collision.

Skill 12.5 Describe the effects of gravitational and nuclear forces in real-life situations through conceptual questions.

Gravitational and the two nuclear forces make up three of the four fundamental forces of nature, with electromagnetic force being the fourth.

Gravitational force is defined as the force of attraction between all masses in the universe. Every object exerts gravitational force on every other object. This force depends on the masses of the objects and the distance between them. The gravitational force between any two masses is given by Newton's law of universal gravitation, which states that the force is inversely proportional to the square of the distance between the masses. Near the surface of the Earth, the acceleration of an object due to gravity is independent of the mass of the object and therefore constant.

It is the gravitational attraction of the Earth that gives weight to objects with mass and causes them to fall to the ground when dropped. Gravitation is responsible for the existence of the objects in our solar system. Without it, the celestial bodies would not be held together. It keeps the Earth and all the other planets in orbit around the sun, keeps the moon in orbit around the Earth, and causes the formation of the tides.

Examples of mechanisms that utilize gravitation to some degree are intravenous drips and water towers where the height difference provides a pressure differential in the liquid. The gravitational potential energy of water is also used to generate hydroelectricity. Pendulum clocks depend upon gravity to regulate time.

There are two types of nuclear forces: strong and weak. The strong force is an interaction that binds protons and neutrons together in atomic nuclei. The strong force only acts on elementary particles directly, but is observed between hadrons (subatomic particles) as the nuclear force.

The weak force is an interaction between elementary particles involving neutrinos or antineutrinos. Its most familiar effects are beta decay and the associated radioactivity.

Skill 12.6 Explain the functions and applications of the instruments and technologies used to study force and motion in everyday life.

The **speedometer** in a car indicates how fast the car is going at any moment (instantaneous speed).

Tachometers measure the speed of rotation of a shaft or disk. In an automobile, this assists the driver in choosing gear settings with a manual transmission. Tachometers are also used in medicine to measure blood flow.

Gravity gradiometers are used in petroleum exploration to determine areas of higher or lower density in the earth's crust.

An **accelerometer** is a device for measuring acceleration. Accelerometers are used along with gyroscopes in inertial guidance systems for rocket programs. One of the most common uses for accelerometers is in airbag deployment systems in automobiles. The accelerometers are used to detect rapid deceleration of the vehicle to determine when a collision has occurred and the severity of the collision. Research is currently being done on the use of accelerometers to improve Global Positioning Systems (GPS). An accelerometer can infer position in places such as tunnels where the GPS cannot detect it. They are incorporated into Tablet PCs to align the screen based upon the direction in which the PC is being held and in many laptop computers to detect falling and protect the data on the hard drive. Accelerometers are incorporated into sports watches to indicate speed and distance (useful to runners).

A **gravimeter** is a device used to measure the local gravitational field. They are much more sensitive than accelerometers. Measurements of the surface gravity of the earth are part of geophysical analysis, which includes the study of earthquakes.

Weighing scales, such as spring scales, are sometimes used to measure force rather than mass or weight.

COMPETENCY 13.0 UNDERSTAND AND APPLY KNOWLEDGE OF ELECTRICITY, MAGNETISM, AND WAVES.

Skill 13.1 Recognize the nature and properties of electricity and magnetism, including static charge, moving charge, basic RC circuits, fields, conductors, and insulators.

An **electric circuit** is a path along which electrons flow. A simple circuit can be created with a dry cell, wire, a bell, or a light bulb. When all are connected, the electrons flow from the negative terminal, through the wire to the device and back to the positive terminal of the dry cell. If there are no breaks in the circuit, the device will work. The circuit is closed. Any break in the flow will create an open circuit and cause the device to shut off.

The device (bell, bulb) is an example of a **load**. A load is a device that uses energy. Suppose that you also add a buzzer so that the bell rings when you press the buzzer button. The buzzer is acting as a **switch**. A switch is a device that opens or closes a circuit. Pressing the buzzer makes the connection complete and the bell rings. When the buzzer is not engaged, the circuit is open and the bell is silent.

A **series circuit** is one where the electrons have only one path along which they can move. When one load in a series circuit goes out, the circuit is open. An example of this is a set of Christmas tree lights that is missing a bulb. None of the bulbs will work.

A **parallel circuit** is one where the electrons have more than one path to move along. If a load goes out in a parallel circuit, the other load will still work because the electrons can still find a way to continue moving along the path.

When an electron goes through a load, it does work and therefore loses some of its energy. The measure of how much energy is lost is called the **potential difference**. The potential difference between two points is the work needed to move a charge from one point to another.

Potential difference is measured in a unit called the volt. **Voltage** is potential difference. The higher the voltage, the more energy the electrons have. This energy is measured by a device called a voltmeter. To use a voltmeter, place it in a circuit parallel with the load you are measuring.

Current is the number of electrons per second that flow past a point in a circuit. Current is measured with a device called an ammeter. To use an ammeter, put it in series with the load you are measuring.

As electrons flow through a wire, they lose potential energy. Some is changed into heat energy because of resistance. **Resistance** is the ability of the material to oppose the flow of electrons through it. All substances have some resistance, even if they are a good conductor such as copper. This resistance is measured in units called **ohms**. A thin wire will have more resistance than a thick one because it will have less room for electrons to travel. In a thicker wire, there will be more possible paths for the electrons to flow. Resistance also depends upon the length of the wire. The longer the wire, the more resistance it will have. Potential difference, resistance, and current form a relationship know as **Ohm's Law**. Current **(I)** is measured in amperes and is equal to potential difference **(V)** divided by resistance **(R)**.

$$I = V / R$$

If you have a wire with resistance of 5 ohms and a potential difference of 75 volts, you can calculate the current by

I = 75 volts / 5 ohms
I = 15 amperes

A current of 10 or more amperes will cause a wire to get hot. 22 amperes is about the maximum for a house circuit. Anything above 25 amperes can start a fire.

Electrostatics is the study of stationary electric charges. A plastic rod that is rubbed with fur or a glass rod that is rubbed with silk will become electrically charged and will attract small pieces of paper. The charge on the plastic rod rubbed with fur is negative and the charge on glass rod rubbed with silk is positive.

Electrically charged objects share these characteristics:

1. Like charges repel one another.
2. Opposite charges attract each other.
3. Charge is conserved. A neutral object has no net change. If the plastic rod and fur are initially neutral, when the rod becomes charged by the fur a negative charge is transferred from the fur to the rod. The net negative charge on the rod is equal to the net positive charge on the fur.

Materials through which electric charges can easily flow are called **conductors**. Metals which are good conductors include silicon and boron. On the other hand, an **insulator** is a material through which electric charges do not move easily, if at all. Examples of insulators would be the nonmetal elements of the periodic table. A simple device used to indicate the existence of a positive or negative charge is called an **electroscope**. An electroscope is made up of a conducting knob and attached to it are very lightweight conducting leaves usually made of foil (gold or aluminum). When a charged object touches the knob, the leaves push away from each other because like charges repel. It is not possible to tell whether if the charge is positive or negative.

Charging by induction:

Touch the knob with a finger while a charged rod is nearby. The electrons will be repulsed and flow out of the electroscope through the hand. If the hand is removed while the charged rod remains close, the electroscope will retain the charge.

When an object is rubbed with a charged rod, the object will take on the same charge as the rod. However, charging by induction gives the object the opposite charge as that of the charged rod.

Grounding charge:

Charge can be removed from an object by connecting it to the earth through a conductor. The removal of static electricity by conduction is called **grounding**.

Skill 13.2   Recognize the nature and properties of mechanical and electromagnetic waves (e.g., frequency, source, medium, spectrum, wave-particle duality).

A mechanical wave can be defined as a disturbance that travels through a medium, moving energy from one place to another. This disturbance is also called an electrical force field. The wave is not capable of transporting energy without a medium, as in vacuum conditions. The medium is the material through which the disturbance is moving and can be thought of as a series of interacting particles. The example of a slinky wave is often used to illustrate the nature of a wave, where pressure exerted on the first coil moves through the remaining coils. A sound wave is also a mechanical wave. The frequency of a wave refers to how often the particles of the medium vibrate when a wave passes through the medium. Frequency is measured in units of cycles/second, waves/second, vibrations/second, or something/second. Another unit for frequency is the Hertz (abbreviated Hz) where 1 Hz is equivalent to 1 cycle/second. The period of a wave is the time required for a particle on a medium to make one complete vibrational cycle. Wave period is measured in units of time such as seconds, hours, days or years, and is NOT synonymous with frequency.

Electromagnetic waves are both electric and magnetic in nature and are capable of traveling through a vacuum. They do not require a medium in order to transport their energy. Light, microwaves, x-rays, and TV and radio transmissions are all kinds of electromagnetic waves. They are all a wavy disturbance that repeats itself over a distance called the wavelength. Electromagnetic waves come in varying sizes and properties, by which they are organized in the electromagnetic spectrum. The electromagnetic spectrum is measured in frequency (f) in hertz and wavelength (λ) in meters. The frequency times the wavelength of every electromagnetic wave equals the speed of light (3.0×10^9 meters/second).

See **Skill 22.1** for more information on the various types of electromagnetic radiation.

Skill 13.3   Describe the effects and applications of electromagnetic forces in real-life situations, including electric power generation, circuit breakers, and brownouts.

Electricity can be used to change the chemical composition of a material. For instance, when electricity is passed through water, it breaks the water down into hydrogen gas and oxygen gas.

Circuit breakers in a home monitor the electric current. If there is an overload, the circuit breaker will create an open circuit, stopping the flow of electricity.

Computers can be made small enough to fit inside a plastic credit card by creating what is known as a solid state device. In this device, electrons flow through solid material such as silicon.

Resistors are used to regulate volume on a television or radio or through a dimmer switch for lights.

A bird can sit on an electrical wire without being electrocuted because the bird and the wire have about the same potential. However, if that same bird would touch two wires at the same time he would not have to worry about flying south next year.

When caught in an electrical storm, a car is a relatively safe place from lightening because of the resistance of the rubber tires. A metal building would not be safe unless there was a lightening rod that would attract the lightening and conduct it into the ground.

A brown-out occurs when there exists a condition of lower than normal power line voltage. This may be short term (minutes to hours) or long term (1/2 day or more). A power line voltage reduction of 8 - 12% is usually considered a Brown-out. Electric utilities may reduce line voltage to Brown-out levels in an effort to manage power generation and distribution. This is most likely to occur on very hot days, when most air conditioning and refrigeration equipment would be operating almost continuously. Even without purposeful intervention from the local utility company, extreme overloads (spikes) could tax the electrical system to the point where a permanent brown-out state could exist over much of the company's distribution network.

Skill 13.4   Analyze and predict the behavior of mechanical and electromagnetic waves under varying physical conditions, including basic optics, color, ray diagrams, and shadows.

The place where one medium ends and another begins is called a **boundary**, and the manner in which a wave behaves when it reaches that boundary is called **boundary behavior**. The following principles apply to boundary behavior in waves:

1) wave speed is always greater in the less dense medium
2) wavelength is always greater in the less dense medium
3) wave frequency is not changed by crossing a boundary
4) the reflected pulse becomes inverted when a wave in a less dense medium is heading towards a boundary with a more dense medium
5) the amplitude of the incident pulse is always greater than the amplitude of the reflected pulse.

For an example, we will use a rope whose left side is less dense, or thinner, than the right side of the rope.

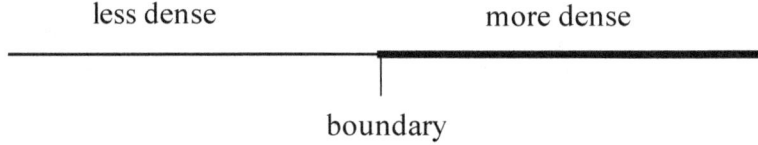

A pulse is introduced on the left end of the rope. This **incident pulse** travels right along the rope towards the boundary between the two thicknesses of rope. When the incident pulse reaches the boundary, two behaviors will occur:

1) Some of the energy will be reflected back to the left side of the boundary. This energy is known as the **reflected pulse**.
2) The rest of the energy will travel into the thicker end of the rope. This energy is referred to as the **transmitted pulse**.

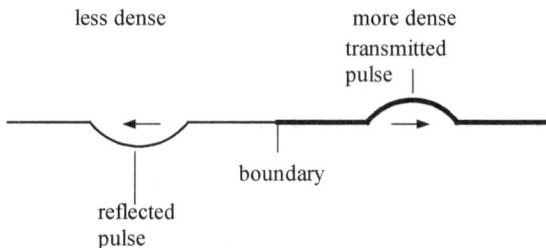

When the incident pulse travels from a denser medium to a less dense medium, the reflected pulse is not inverted.

Reflection occurs when waves bounce off a barrier. The **law of reflection** states that when a ray of light reflects off a surface, the angle of incidence is equal to the angle of reflection.

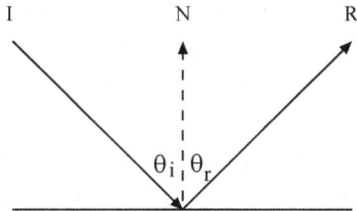

Line I represents the **incident ray**, the ray of light striking the surface. Line R is the **reflected ray**, the ray of light reflected off the surface. Line N is known as the **normal line**. It is a perpendicular line at the point of incidence that divides the angle between the incident ray and the reflected ray into two equal rays. The angle between the incident ray and the normal line is called the **angle of incidence**; the angle between the reflected ray and the normal line is called the **angle of reflection**.

Waves passing from one medium into another will undergo **refraction**, or bending. Accompanying this bending are a change in both speed and the wavelength of the waves.

In this example, light waves traveling through the air will pass through glass.

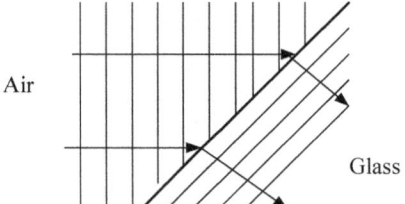

Refraction occurs only at the boundary. Once the wavefront passes across the boundary, it travels in a straight line.

Diffraction involves a change in direction of waves as they pass through an opening or around an obstacle in their path.

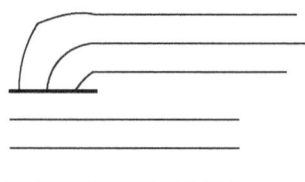

The amount of diffraction depends upon the wavelength. The amount of diffraction increases with increasing wavelength and decreases with decreasing wavelength. Sound and water waves exhibit this ability.

When we refer to light, we are usually talking about a type of electromagnetic wave that stimulates the retina of the eye, or visible light. Each individual wavelength within the spectrum of visible light represents a particular **color**. When a particular wavelength strikes the retina, we perceive that color. Visible light is sometimes referred to as ROYGBIV (red, orange, yellow, green, blue, indigo, violet). The visible light spectrum ranges from red (the longest wavelength) to violet (the shortest wavelength) with a range of wavelengths in between. If all the wavelengths strike your eye at the same time, you will see white. Conversely, when no wavelengths strike your eye, you perceive black.

A **shadow** results from the inability of light waves to diffract as sound and water waves can. An obstacle in the way of the light waves blocks the light waves, thereby creating a shadow.

TEACHER CERTIFICATION STUDY GUIDE

SUBAREA IV.   **EARTH SYSTEMS AND THE UNIVERSE**

COMPETENCY 14.0 UNDERSTAND AND APPLY KNOWLEDGE OF EARTH'S LAND, WATER, AND ATMOSPHERIC SYSTEMS AND THE HISTORY OF EARTH

Skill 14.1   Identify the structure and composition of Earth's land, water, and atmospheric systems and how they affect weather, erosion, fresh water, and soil.

Water is recycled throughout ecosystems. Just two percent of all the available water is fixed and held in ice or the bodies of organisms. Available water includes surface water (lakes, ocean, and rivers) and ground water (aquifers, wells). 96% of all available water is from ground water. Water is recycled through the processes of evaporation and precipitation. The water present now is the water that has been here since our atmosphere formed.

When water falls from the atmosphere it can have an erosive property. This can happen from impact alone but also from acid rain. Erosion is known for its destructive properties, but in an indirect way, it also builds by bringing materials to new locations. **Erosion** is the inclusion and transportation of surface materials by another moveable material, usually water, wind, or ice. The most important cause of erosion is running water. Streams, rivers, and tides are constantly at work removing weathered fragments of bedrock and carrying them away from their original location. A stream erodes bedrock by the grinding action of the sand, pebbles and other rock fragments. This grinding against each other is called abrasion. Streams also erode rocks by dissolving or absorbing their minerals. Limestone and marble are readily dissolved by streams.

The breaking down of rocks at or near to the earth's surface is known as **weathering**. Weathering breaks down these rocks into smaller and smaller pieces. There are two types of weathering: physical weathering and chemical weathering.

Physical weathering is the process by which rocks are broken down into smaller fragments without undergoing any change in chemical composition. Physical weathering is mainly caused by the freezing of water, the expansion of rock, and the activities of plants and animals.

Frost wedging is the cycle of daytime thawing and refreezing at night. This cycle causes large rock masses, especially the rocks exposed on mountain tops, to be broken into smaller pieces.

The peeling away of the outer layers from a rock is called exfoliation. Rounded mountain tops are called exfoliation domes and have been formed in this way.

Chemical weathering is the breaking down of rocks through changes in their chemical composition. An example would be the change of feldspar in granite to clay. Water, oxygen, and carbon dioxide are the main agents of chemical weathering. When water and carbon dioxide combine chemically, they produce a weak acid that breaks down rocks. In addition, acidic substances from factories and car exhausts dissolve in rain water forming **acid rain.** Acid rain forms predominantly from pollutant oxides in the air (usually nitrogen-based NO_x or sulfur-based SO_x), which become hydrated into their acids (nitric or sulfuric acid). When the rain falls into stone, the acids can react with metallic compounds and gradually wear the stone away.

Skill 14.2 Recognize the scope of geologic time and the continuing physical changes of Earth through time.

Geological time is divided into periods depending on the kind of life that existed at that time. These periods are grouped together into eras. The history of the Earth is calculated by studying the ages of the various layers of sedimentary rock.

Era	Period	Time	Characteristics
Cenozoic	Quaternary	1.6 million years ago to the present.	The Ice Age occurred, and human beings evolved.
	Tertiary	65-1.64 million years ago.	Mammals and birds evolved to replace the great reptiles and dinosaurs that had just become extinct. Forests gave way to grasslands, and the climate become cooler.
Mesozoic	Cretaceous	135-65 million years ago.	Reptiles and dinosaurs roamed the Earth. Most of the modern continents had split away from the large landmass, Pangaea, and many were flooded by shallow chalk seas.
	Jurassic	350-135 million years ago.	Reptiles were beginning to evolve. Pangaea started to break up. Deserts gave way to forests and swamps.
	Triassic		

Paleozoic	Permian	355-250 million years ago.	Continents came together to form one big landmass, Pangaea. Forests (that formed today's coal) grew on deltas around the new mountains, and deserts formed.
	Carboniferous		
	Devonian	410-355 million years ago.	Continents started moving toward each other. The first land animals, such as insects and amphibians, existed. Many fish swam in the seas.
	Silurian	510-410 million years ago.	Sea life flourished, and the first fish evolved. The earliest land plants began to grow around shorelines and estuaries.
	Ordovician		
	Cambrian	570-510 million years ago.	No life on land, but all kinds of sea animals existed.
Precambrian	Proterozoic	Beginning of the Earth to 570 million years ago (seven-eighths of the Earth's history).	Some sort of life existed.
	Archaean		No life.

Skill 14.3   Evaluate scientific theories about Earth's origin and history and how these theories explain contemporary living systems.

The dominant scientific theory about the origin of the Universe, and consequently the Earth, is the **Big Bang Theory**. According to this theory, an atom exploded about 10 to 20 billion years ago throwing matter in all directions. Although this theory has never been proven, and probably never will be, it is supported by the fact that distant galaxies in every direction are moving away from us at great speeds.

Earth, itself, is believed to have been created 4.5 billion years ago as a solidified cloud of gases and dust left over from the creation of the sun. As millions of years passed, radioactive decay released energy that melted some of Earth's components. Over time, the heavier components sank to the center of the Earth and accumulated into the core. As the Earth cooled, a crust formed with natural depressions. Water rising from the interior of the Earth filled these depressions and formed the oceans. Slowly, the Earth acquired the appearance it has today.

The **Heterotroph Hypothesis** supposes that life on Earth evolved from **heterotrophs**, the first cells. According to this hypothesis, life began on Earth about 3.5 billion years ago. Scientists have shown that the basic molecules of life formed from lightning, ultraviolet light, and radioactivity. Over time, these molecules became more complex and developed metabolic processes, thereby becoming heterotrophs. Heterotrophs could not produce their own food and fed off organic materials. However, they released carbon dioxide which allowed for the evolution of **autotrophs**, which could produce their own food through photosynthesis. The autotrophs and heterotrophs became the dominant life forms and evolved into the diverse forms of life we see today.

Proponents of creationism believe that the species we currently have were created as recounted in the book of Genesis in the Bible. This retelling asserts that God created all life about 6,000 years ago in one mass creation event. However, scientific evidence casts doubt on creationism.

Evolution

The most significant evidence to support the history of evolution is fossils, which have been used to construct a fossil record. Fossils give clues as to the structure of organisms and the times at which they existed. However, there are limitations to the study of fossils, which leave huge gaps in the fossil record.

Scientists also try to relate two organisms by comparing their internal and external structures. This is called **comparative anatomy**. Comparative anatomy categorizes anatomical structures as **homologous** (features in different species that point to a common ancestor), **analogous** (structures that have superficial similarities because of similar functions, but do not point to a common ancestor), and **vestigial** (structures that have no modern function, indicating that different species diverged and evolved). Through the study of **comparative embryology**, homologous structures that do not appear in mature organisms may be found between different species in their embryological development.

There have been two basic **theories of evolution: Lamarck's and Darwin's**. Lamarck's theory that proposed that an organism can change its structure through use or disuse and that acquired traits can be inherited has been disproved.

Darwin's theory of **natural selection** is the basis of all evolutionary theory. His theory has four basic points:

1. Each species produces more offspring than can survive.
2. The individual organisms that make up a larger population are born with certain variations.
3. The overabundance of offspring creates competition for survival among individual organisms (**survival of the fittest**).
4. Variations are passed down from parent to offspring.
5.

Points 2 and 4 form the genetic basis for evolution.

New species develop from two types of evolution: divergent and convergent. **Divergent evolution**, also known as **speciation**, is the divergence of a new species from a previous form of that species. There are two main ways in which speciation may occur: **allopatric speciation** (resulting from geographical isolation so that species cannot interbreed) and **adaptive radiation** (creation of several new species from a single parent species). **Convergent evolution** is a process whereby different species develop similar traits from inhabiting similar environments, facing similar selection pressures, and/or use parts of their bodies for similar functions. This type of evolution is only superficial. It can never result in two species being able to interbreed.

Skill 14.4 Recognize the interrelationships between living organisms and Earth's resources and evaluate the uses of Earth's resources.

The region of the Earth and its atmosphere in which living things are found is known as the **biosphere**. The biosphere is made up of distinct areas called **ecosystems**, each of which has its own characteristic climate, soils, and communities of plants and animals.
The most important nonliving factors affecting an ecosystem are the chemical cycles, the water cycle, oxygen, sunlight, and the soil. The two basic chemical cycles are the carbon cycle and the nitrogen cycle. They involve the passage of these elements between the organisms and the environment.

In the **carbon cycle**, animals and plants use carbon dioxide from the air to produce glucose, which they use in respiration and other life processes. Animals consume plants, use what they can of the carbon matter and excrete the rest as waste. This waste decays into carbon dioxide. During respiration, plants and animals release carbon dioxide back into the air. The carbon used by plants and animals stays in their bodies until death, after which decay sends the organic compounds back into the Earth and carbon dioxide back into the air.

Nitrogen found in the atmosphere is generally unusable by living organisms. In the **nitrogen cycle,** nitrogen-fixing bacteria in the soil and/or the roots of legumes transform the inert nitrogen into compounds. Plants take these compounds, synthesize the twenty amino acids found in nature, and turn them into plant proteins. Animals can only synthesize eight of the amino acids. They eat the plants to produce protein from the plant's materials. Animals and plants give off nitrogen waste and death products in the form of ammonia. The ammonia will either be transformed into nitrites and nitrates by bacteria and reenter the cycle when they are taken up by plants, or be broken down by bacteria to produce inert nitrogen to be released back into the air.

Most of the Earth's water is found in the oceans and lakes. Through the **water cycle**, water evaporates into the atmosphere and condenses into clouds. Water then falls to the Earth in the form of precipitation, returning to the oceans and lakes on falling on land. Water on the land may return to the oceans and lakes as runoff or seep from the soil as groundwater.

The amount of **oxygen** available in a particular location may create competition. Oxygen is readily available to animals on land; but in order for it to be available to aquatic organisms, it must be dissolved in water.

Sunlight is also important to most organisms. Organisms on land compete for sunlight, but sunlight does not reach into the lowest depths of the ocean. Organisms in these regions must find another means of producing food.

The type of **soil** found in a particular ecosystem determines what species can live in that ecosystem.

COMPETENCY 15.0 UNDERSTAND AND APPLY KNOWLEDGE OF THE DYNAMIC NATURE OF EARTH

Skill 15.1 Analyze and explain large-scale dynamic forces, events, and processes that affect Earth's land, water, and atmospheric systems, including conceptual questions about plate tectonics, El Nino, drought, and climatic shifts.

El Niño refers to a sequence of changes in the ocean and atmospheric circulation across the Pacific Ocean. The water around the equator is unusually hot every two to seven years. Trade winds normally blowing east to west across the equatorial latitudes, piling warm water into the western Pacific. A huge mass of heavy thunderstorms usually forms in the area and produce vast currents of rising air that displace heat poleward. This helps create the strong mid-latitude jet streams. The world's climate patterns are disrupted by this change in location of the massive cluster of thunderstorms. The West coast of America experienced a wet winter. Sacramento, California recorded 103 days of rain.

Air masses moving toward or away from the Earth's surface are called air currents. Air moving parallel to Earth's surface is called **wind**. Weather conditions are generated by winds and air currents carrying large amounts of heat and moisture from one part of the atmosphere to another. Wind speeds are measured by instruments called anemometers.

The wind belts in each hemisphere consist of convection cells that encircle Earth like belts. There are three major wind belts on Earth (1) trade winds (2) prevailing westerlies, and (3) polar easterlies. Wind belt formation depends on the differences in air pressures that develop in the doldrums, the horse latitudes, and the polar regions. The Doldrums surround the equator. Within this belt heated air usually rises straight up into Earth's atmosphere. The Horse latitudes are regions of high barometric pressure with calm and light winds and the Polar regions contain cold dense air that sinks to the earth's surface

Winds caused by local temperature changes include sea breezes, and land breezes.

Sea breezes are caused by the unequal heating of the land and an adjacent, large body of water. Land heats up faster than water. The movement of cool ocean air toward the land is called a sea breeze. Sea breezes usually begin blowing about mid-morning; ending about sunset.

A breeze that blows from the land to the ocean or a large lake is called a **land breeze.**

Monsoons are huge wind systems that cover large geographic areas and that reverse direction seasonally. The monsoons of India and Asia are examples of these seasonal winds. They alternate wet and dry seasons. As denser cooler air over the ocean moves inland, a steady seasonal wind called a summer or wet monsoon is produced.

Cloud types:

Cirrus clouds - White and feathery high in sky

Cumulus – thick, white, fluffy

Stratus – layers of clouds cover most of the sky

Nimbus – heavy, dark clouds that represent thunderstorm clouds

Variation on the clouds mentioned above:

Cumulo-nimbus

Strato-nimbus

The air temperature at which water vapor begins to condense is called the **dew point.**

Relative humidity is the actual amount of water vapor in a certain volume of air compared to the maximum amount of water vapor this air could hold at a given temperature.

Skill 15.2   Identify and explain Earth processes and cycles and cite examples in real-life situations, including conceptual questions on rock cycles, volcanism, and plate tectonics.

Data obtained from many sources led scientists to develop the theory of plate tectonics. This theory is the most current model that explains not only the movement of the continents, but also the changes in the earth's crust caused by internal forces.

Plates are rigid blocks of earth's crust and upper mantle. These solid blocks make up the lithosphere. The earth's lithosphere is broken into nine large sections and several small ones. These moving slabs are called plates. The major plates are named after the continents they are "transporting."

The plates float on and move with a layer of hot, plastic-like rock in the upper mantle. Geologists believe that the heat currents circulating within the mantle cause this plastic zone of rock to slowly flow, carrying along the overlying crustal plates.

Movement of these crustal plates creates areas where the plates diverge as well as areas where the plates converge. A major area of divergence is located in the Mid-Atlantic. Currents of hot mantle rock rise and separate at this point of divergence creating new oceanic crust at the rate of 2 to 10 centimeters per year. Convergence is when the oceanic crust collides with either another oceanic plate or a continental plate. The oceanic crust sinks forming an enormous trench and generating volcanic activity. Convergence also includes continent to continent plate collisions. When two plates slide past one another a transform fault is created.

These movements produce many major features of the earth's surface, such as mountain ranges, volcanoes, and earthquake zones. Most of these features are located at plate boundaries, where the plates interact by spreading apart, pressing together, or sliding past each other. These movements are very slow, averaging only a few centimeters each year.

Boundaries form between spreading plates where the crust is forced apart in a process called rifting. Rifting generally occurs at mid-ocean ridges. Rifting can also take place within a continent, splitting the continent into smaller landmasses that drift away from each other, thereby forming an ocean basin between them. The Red Sea is a product of rifting. As the seafloor spreading takes place, new material is added to the inner edges of the separating plates. In this way the plates grow larger, and the ocean basin widens. This is the process that broke up the super continent Pangaea and created the Atlantic Ocean.

Boundaries between plates that are colliding are zones of intense crustal activity. When a plate of ocean crust collides with a plate of continental crust, the more dense oceanic plate slides under the lighter continental plate and plunges into the mantle. This process is called **subduction**, and the site where it takes place is called a subduction zone. A subduction zone is usually seen on the sea-floor as a deep depression called a trench.

The crustal movement that is identified by plates sliding sideways past each other produces a plate boundary characterized by major faults that are capable of unleashing powerful earth-quakes. The San Andreas Fault forms such a boundary between the Pacific Plate and the North American Plate.

Orogeny is the term given to natural mountain building.

A mountain is terrain that has been raised high above the surrounding landscape by volcanic action, or some form of tectonic plate collisions. The plate collisions could be intercontinental or ocean floor collisions with a continental crust (subduction). The physical composition of mountains would include igneous, metamorphic, or sedimentary rocks; some may have rock layers that are tilted or distorted by plate collision forces.

There are many different types of mountains. The physical attributes of a mountain range depends upon the angle at which plate movement thrust layers of rock to the surface. Many mountains (Adirondacks, Southern Rockies) were formed along high angle faults.

Folded mountains (Alps, Himalayas) are produced by the folding of rock layers during their formation. The Himalayas are the highest mountains in the world and contain Mount Everest, which rises almost 9 km above sea level. The Himalayas were formed when India collided with Asia. The movement that created this collision is still in process at the rate of a few centimeters per year.

Fault-block mountains (Utah, Arizona, and New Mexico) are created when plate movement produces tension forces instead of compression forces. The area under tension produces normal faults and rock along these faults is displaced upward.

Dome mountains are formed as magma tries to push up through the crust but fails to break the surface. Dome mountains resemble a huge blister on the earth's surface.

Upwarped mountains (Black Hills of South Dakota) are created in association with a broad arching of the crust. They can also be formed by rock thrust upward along high angle faults.

Faults are categorized on the basis of the relative movement between the blocks on both sides of the fault plane. The movement can be horizontal, vertical or oblique.

A dip-slip fault occurs when the movement of the plates is vertical and opposite. The displacement is in the direction of the inclination, or dip, of the fault. Dip-slip faults are classified as normal faults when the rock above the fault plane moves down relative to the rock below.

Reverse faults are created when the rock above the fault plane moves up relative to the rock below. Reverse faults having a very low angle to the horizontal are also referred to as thrust faults.

Faults in which the dominant displacement is horizontal movement along the trend or strike (length) of the fault are called **strike-slip faults**. When a large strike-slip fault is associated with plate boundaries it is called a **transform fault**. The San Andreas Fault in California is a well-known transform fault.

Faults that have both vertical and horizontal movement are called **oblique-slip faults.**

Volcanism is the term given to the movement of magma through the crust and its emergence as lava onto the earth's surface. Volcanic mountains are built up by successive deposits of volcanic materials.

An active volcano is one that is presently erupting or building to an eruption. A dormant volcano is one that is between eruptions but still shows signs of internal activity that might lead to an eruption in the future. An extinct volcano is said to be no longer capable of erupting. Most of the world's active volcanoes are found along the rim of the Pacific Ocean, which is also a major earthquake zone. This curving belt of active faults and volcanoes is often called the Ring of Fire. The world's best known volcanic mountains include: Mount Etna in Italy and Mount Kilimanjaro in Africa. The Hawaiian Islands are actually the tops of a chain of volcanic mountains that rise from the ocean floor.

There are three types of volcanic mountains: shield volcanoes, cinder cones, and composite volcanoes.

Shield Volcanoes are associated with quiet eruptions. Lava emerges from the vent or opening in the crater and flows freely out over the earth's surface until it cools and hardens into a layer of igneous rock. A repeated lava flow builds this type of volcano into the largest volcanic mountain. Mauna Loa found in Hawaii, is the largest volcano on earth.

Cinder Cone Volcanoes are associated with explosive eruptions as lava is hurled high into the air in a spray of droplets of various sizes. These droplets cool and harden into cinders and particles of ash before falling to the ground. The ash and cinder pile up around the vent to form a steep, cone-shaped hill called the cinder cone. Cinder cone volcanoes are relatively small but may form quite rapidly.

Composite Volcanoes are described as being built by both lava flows and layers of ash and cinders. Mount Fuji in Japan, Mount St. Helens in Washington, USA, and Mount Vesuvius in Italy are all famous composite volcanoes.

When lava cools, **igneous rock** is formed. This formation can occur either above ground or below ground.

Intrusive rock includes any igneous rock that was formed below the earth's surface. Batholiths are the largest structures of intrusive type rock and are composed of near granite materials; they are at the core of the Sierra Nevada Mountains.

Extrusive rock includes any igneous rock that was formed at the earth's surface.

Dikes are old lava tubes formed when magma entered a vertical fracture and hardened. Sometimes magma squeezes between two rock layers and hardens into a thin horizontal sheet called a **sill**. A **laccolith** is formed in much the same way as a sill, but the magma that creates a laccolith is very thick and does not flow easily. It pools and forces the overlying strata up, creating an obvious surface dome.

A **caldera** is normally formed by the collapse of the top of a volcano. This collapse can be caused by a massive explosion that destroys the cone and empties most, if not all, of the magma chamber below the volcano. The cone collapses into the empty magma chamber forming a caldera.

An inactive volcano may have magma solidified in its pipe. This structure, called a volcanic neck, is resistant to erosion and today may be the only visible evidence of the past presence of an active volcano.

Skill 15.3 **Analyze the transfer of energy within and among Earth's land, water, and atmospheric systems, including the identification of energy sources of volcanoes, hurricanes, thunderstorms, and tornadoes.**

A **thunderstorm** is a brief, local storm produced by the rapid upward movement of warm, moist air within a cumulo-nimbus cloud. Thunderstorms always produce lightning and thunder, accompanied by strong wind gusts and heavy rain or hail.

A severe storm with swirling winds that may reach speeds of hundreds of km per hour is called a **tornado**. Such a storm is also referred to as a "twister". The sky is covered by large cumulo-nimbus clouds and violent thunderstorms; a funnel-shaped swirling cloud may extend downward from a cumulonimbus cloud and reach the ground. Tornadoes are narrow storms that leave a narrow path of destruction on the ground.

A swirling, funnel-shaped cloud that **extends** downward and touches a body of water is called a **waterspout**.

Hurricanes are storms that develop when warm, moist air carried by trade winds rotates around a low-pressure "eye". A large, rotating, low-pressure system accompanied by heavy precipitation and strong winds is called a tropical cyclone or is better known as a hurricane. In the Pacific region, a hurricane is called a typhoon.

Storms that occur only in the winter are known as blizzards or ice storms. A **blizzard** is a storm with strong winds, blowing snow and frigid temperatures. An **ice storm** consists of falling rain that freezes when it strikes the ground, covering everything with a layer of ice.

Skill 15.4 **Explain the functions and applications of the instruments and technologies used to study the earth sciences, including seismographs, barometers, and satellite systems.**

Satellites have improved our ability to communicate and transmit radio and television signals. Navigational abilities have been greatly improved through the use of satellite signals. Sonar uses sound waves to locate objects and is especially useful underwater. The sound waves bounce off the object and are used to assist in location. **Seismographs** record vibrations in the earth and allow us to measure earthquake activity. Common instruments for that forecasting weather include the **aneroid barometer** and the **mercury barometer,** which both measure air pressure. In the aneroid barometer, the air exerts varying pressures on a metal diaphragm that will then read air pressure. The mercury barometer operates when atmospheric pressure pushes on a pool of mercury in a glass tube. The higher the pressure, the higher up the tube mercury will rise.

Relative humidity is measured by two kinds of additional weather instruments, the psychrometer and the hair gygrometer.

TEACHER CERTIFICATION STUDY GUIDE

COMPETENCY 16.0 UNDERSTAND AND APPLY KNOWLEDGE OF OBJECTS IN THE UNIVERSE AND THEIR DYNAMIC INTERACTIONS.

Skill 16.1   Describe and explain the relative and apparent motions of the sun, the moon, stars, and planets in the sky.

Until the summer of 2006, there were nine recognized planets in our solar system: Mercury, Venus, Earth, Mars, Jupiter, Saturn, Uranus, Neptune, and Pluto. These nine planets are divided into two groups based on distance from the sun. The inner planets include: Mercury, Venus, Earth, and Mars. The outer planets include: Jupiter, Saturn, Uranus, Neptune and Pluto. Pluto's status as a planet is being reconsidered.

Mercury -- the closest planet to the sun. Its surface has craters and rocks. The atmosphere is composed of hydrogen, helium and sodium. Mercury was named after the Roman messenger god.

Venus -- has a slow rotation when compared to Earth. Venus and Uranus rotate in opposite directions from the other planets. This opposite rotation is called retrograde rotation. The surface of Venus is not visible due to the extensive cloud cover. The atmosphere is composed mostly of carbon dioxide. Sulfuric acid droplet in the dense cloud cover gives Venus a yellow appearance. Venus has a greater greenhouse effect than observed on Earth. The dense clouds Combined with carbon dioxide traps heat. Venus was named after the Roman goddess of love.

Earth -- considered a water planet with 70% of its surface covered with water. Gravity holds the masses of water in place. The different temperatures observed on earth allows for the different states of water to exist; solid, liquid or gas. The atmosphere is composed mainly of oxygen and nitrogen. Earth is the only planet that is known to support life.

Mars -- the surface of Mars contains numerous craters, active and extinct volcanoes, ridges and valleys with extremely deep fractures. Iron oxide found in the dusty soil makes the surface seem rust colored and the skies seem pink in color. The atmosphere is composed of carbon dioxide, nitrogen, argon, oxygen and water vapor. Mars has polar regions with ice caps composed of water. Mars has two satellites. Mars was named after the Roman war god.

Jupiter -- largest planet in the solar system. Jupiter has 16 moons. The atmosphere is composed of hydrogen, helium, methane and ammonia. There are white colored bands of clouds indicating rising gas and dark colored bands of clouds indicating descending gases, caused by heat resulting from the energy of Jupiter's core. Jupiter has a Great Red Spot that is thought to be a hurricane type cloud. Jupiter has a strong magnetic field.

Saturn -- the second largest planet in the solar system. Saturn has beautiful rings of ice and rock and dust particles circling it. Saturn's atmosphere is composed of hydrogen, helium, methane, and ammonia. Saturn has 20 plus satellites. Saturn was named after the Roman god of agriculture.

Uranus -- the second largest planet in the solar system with retrograde revolution. Uranus a gaseous planet and it has 10 dark rings and 15 satellites. Its atmosphere is composed of hydrogen, helium, and methane. Uranus was named after the Greek god of the heavens.

Neptune -- another gaseous planet with an atmosphere consisting of hydrogen, helium, and methane. Neptune has 3 rings and 2 satellites. Neptune was named after the Roman sea god that its atmosphere has the same color of the seas.

Pluto -- considered the smallest planet in the solar system. Pluto's atmosphere probably contains methane, ammonia, and frozen water. Pluto has 1 satellite. Pluto revolves around the sun every 250 years. Pluto was named after the Roman god of the underworld.

Skill 16.2   Recognize properties of objects (e.g., comets, asteroids) within the solar system and their dynamic interactions.

Astronomers believe that the rocky fragments that may have been the remains of the birth of the solar system that never formed into a planet. **Asteroids** are found in the region between Mars and Jupiter.

Comets are masses of frozen gases, cosmic dust, and small rocky particles. Astronomers think that most comets originate in a dense comet cloud beyond Pluto. Comet consists of a nucleus, a coma, and a tail. A comet's tail always points away from the sun. The most famous comet, **Halley's Comet**, is named after the person whom first discovered it in 240 B.C. It returns to the skies near earth every 75 to 76 years.

Meteoroids are composed of particles of rock and metal of various sizes. When a meteoroid travels through the earth's atmosphere, friction causes its surface to heat up and it begins to burn. The burning meteoroid falling through the earth's atmosphere is now called a **meteor** or also known as a "shooting star."

Meteorites are meteors that strike the earth's surface. A physical example of the impact of the meteorite on the earth's surface can be seen in Arizona, The Barringer Crater is a huge Meteor Crater. There many other such meteor craters found throughout the world.

TEACHER CERTIFICATION STUDY GUIDE

Skill 16.3 Recognize the types, properties, and dynamics of objects external to the solar system (e.g., black holes, supernovas, galaxies).

Astronomers use groups or patterns of stars called **constellations** as reference points to locate other stars in the sky. Familiar constellations include: Ursa Major (also known as the big bear) and Ursa Minor (known as the little bear). Within the Ursa Major, the smaller constellation, The Big Dipper is found. Within the Ursa Minor, the smaller constellation, The Little Dipper is found.

Different constellations appear as the earth continues its revolution around the sun with the seasonable changes.

Magnitude stars are 21 of the brightest stars that can be seen from earth, these are the first stars noticed at night. In the Northern Hemisphere there are 15 commonly observed first magnitude stars.

A vast collection of stars is defined as **galaxies**. Galaxies are classified as irregular, elliptical, and spiral. An irregular galaxy has no real structured appearance; most are in their early stages of life. An elliptical galaxy is smooth ellipses, containing little dust and gas, but composed of millions or trillion stars. Spiral galaxies are disk-shaped and have extending arms that rotate around its dense center. Earth's galaxy is found in the Milky Way and it is a spiral galaxy.

A **pulsar** is defined as a variable radio source that emits signals in very short, regular bursts; believed to be a rotating neutron star.

A **quasar** is defined as an object that photographs like a star but has an extremely large redshift and a variable energy output; believed to be the active core of a very distant galaxy.

Black holes are defined as an object that has collapsed to such a degree that light can not escape from its surface; light is trapped by the intense gravitational field.

TEACHER CERTIFICATION STUDY GUIDE

COMPETENCY 17.0 UNDERSTAND AND APPLY KNOWLEDGE OF THE ORIGINS OF AND CHANGES IN THE UNIVERSE.

Skill 17.1 Identify scientific theories dealing with the origin of the universe (e.g., big bang).

Two main theories to explain the origins of the universe include: (1) **The Big Bang Theory** and (2) **The Steady-State Theory.**

The Big Bang Theory has been widely accepted by many astronomers. It states that the universe originated from a magnificent explosion spreading mass, matter and energy into space. The galaxies formed from this material as it cooled during the next half-billion years.

The Steady-State Theory is the least accepted theory. It states that the universe is a continuously being renewed. Galaxies move outward and new galaxies replace the older galaxies. Astronomers have not found any evidence to prove this theory.

The future of the universe is hypothesized with the Oscillating Universe Hypothesis. It states that the universe will oscillate or expand and contract. Galaxies will move away from one another and will in time slow down and stop. Then a gradual moving toward each other will again activate the explosion or The Big Bang theory.

Skill 17.2 Analyze evidence relating to the origin and physical evolution of the universe (e.g., microwave background radiation, expansion).

Cosmic microwave background radiation (CMBR) is the oldest light we can see. It is a snapshot of how the universe looked in its early beginnings. First discovered in 1964, CMBR is composed of photons which we can see because of the atoms that formed when the universe cooled to 3000 K. Prior to that, after the Big Bang, the universe was so hot that the photons were scattered all over the universe, making the universe opaque. The atoms caused the photons to scatter less and the universe to become transparent to radiation. Since cooling to 3000K, the universe has continued to expand and cool.

COBE, launched in 1989, was the first mission to explore slight fluctuations in the background. WMAP, launched in 2001, took a clearer picture of the universe, providing evidence to support the Big Bang Theory and add details to the early conditions of the universe. Based upon this more recent data, scientists believe the universe is about 13.7 billion years old and that there was a period of rapid expansion right after the Big Bang. They have also learned that there were early variations in the density of matter resulting in the formation of the galaxies, the geometry of the universe is flat, and the universe will continue to expand forever.

Skill 17.3   Compare the physical and chemical processes involved in the life cycles of objects within galaxies.

Scientists believe that **stars** form when compression waves traveling through clouds of gas create knots of gas in the clouds. The force of gravity within these denser areas then attracts gas particles. As the knot grows, the force increases and attracts more gas particles, eventually forming a large sphere of compressed gas with internal temperatures reaching a few million degrees C. At these temperatures, the gases in the knot become so hot that nuclear fusion of hydrogen to form helium takes place, creating large amounts of nuclear energy and forming a new star. Pressure from the radiation of these new stars causes more knots to form in the gas cloud, initiating the process of creating more stars.

Scientists theorize that **planets** form from gas and dust surrounding young stars. As the density of the forming star increases, this gas and dust slowly condenses into a spinning disk. The denser areas of the disk develop a gravitational force which attracts more dust and gas as the disk orbits the star. Over millions of years, these dense areas consolidate and grow, forming planets. In the case of the Sun, the larger icy fragments surrounding it attracted more gas and dust forming the more massive planets such as Jupiter and Saturn. These larger planets developed gravitational forces great enough to attract hydrogen and helium atoms, turning them into gas giants. The smaller planets, such as Earth, could not attract these atoms and became mainly rocky.

It is believed that **black holes** form as stars evolve. As the nuclear fuels are used up in the core of a star, the pressure associated with the production of these fuels no longer exists to resist contraction of the core. Two new types of pressure, electron and neutron, arise. However, if the star is more than about five times as massive as the Sun, neither pressure will prevent the star from collapsing into a black hole.

When the universe was forming, most of the material became concentrated in the planets and moons. There were however many small, rocky objects called **planetesimals** that also formed from the gas and dust. These planetesimals include **comets** and **asteroids**. A large cloud of comets, known as the Oort cloud, exists beyond Pluto. A change in the gravitational pull of our galaxy may disturb the orbit of a comet causing it to fall toward the Sun. The ice in the comet turns into vapor, releasing dust from the body. Gas and dust then form the tail of the comet.

In the early life of the solar system, some of the planetesimals came together more toward the center of the solar system. The gravitational pull of Jupiter prevented these planetesimals from developing into full planets. They broke up into thousands of minor planets, known as asteroids.

It is believed that **black holes** form as stars evolve. As the nuclear fuels are used up in the core of a star, the pressure associated with the production of these fuels no longer exists to resist contraction of the core. Two new types of pressure, electron and neutron, arise. However, if the star is more than about five times as massive as the Sun, neither pressure will prevent the star from collapsing into a black hole.

When the universe was forming, most of the material became concentrated in the planets and moons. There were however many small, rocky objects called **planetesimals** that also formed from the gas and dust. These planetesimals include **comets** and **asteroids**. A large cloud of comets, known as the Oort cloud, exists beyond Pluto. A change in the gravitational pull of our galaxy may disturb the orbit of a comet causing it to fall toward the Sun. The ice in the comet turns into vapor, releasing dust from the body. Gas and dust then form the tail of the comet.

In the early life of the solar system, some of the planetesimals came together more toward the center of the solar system. The gravitational pull of Jupiter prevented these planetesimals from developing into full planets. They broke up into thousands of minor planets, known as asteroids.

Skill 17.4 **Explain the functions and applications of the instruments, technologies, and tools used in the study of the space sciences, including the relative advantages and disadvantages of Earth-based versus space-based instruments and optical versus nonoptical instruments.**

Types of telescopes used in the study of the space sciences include optical, radio, infrared, ultraviolet, x-ray, and gamma-ray. Optical telescopes work by collecting and magnifying visible light that is given off by stars or reflected from the surfaces of the planets. However, stars also give off other types of electromagnetic radiation, including radio waves, microwaves, infrared light, ultraviolet light, X rays, and gamma rays. Therefore, specific types of non-optical instruments have been developed to collect information about the universe through these other types of electromagnetic waves.

Many of the telescopes used by astronomers are earth-based, located in observatories around the world. However, only radio waves, visible light, and some infrared radiation can penetrate our atmosphere to reach the earth's surface. Therefore, scientists have launched telescopes into space, where the instruments can collect other types of electromagnetic waves. Space probes are also able to gather information from distant parts of the solar system. In addition to telescopes, scientists construct mathematical models and computer simulations to form a scientific account of events in the universe. These models and simulations are built using evidence from many sources, including the information gathered through telescopes and space probes.

TEACHER CERTIFICATION STUDY GUIDE

SUBAREA V **PHYSICS SKILLS, MOTION, FORCES, AND WAVES**

COMPETENCY 18.0 UNDERSTAND AND APPLY THE KNOWLEDGE AND SKILLS NEEDED TO PRACTICE PHYSICS AND UNDERSTAND THE BROAD APPLICABILITY OF ITS PRINCIPLES TO REAL-WORLD SITUATIONS.

Skill 18.1 Demonstrate knowledge of the safe and proper use of equipment and materials commonly used in physics classrooms and laboratories

Safety is a learned behavior and must be incorporated into instructional plans. Measures of prevention and procedures for dealing with emergencies in hazardous situations have to be in place and readily available for reference. Copies of these must be given to all people concerned, such as administrators and students.

The single most important aspect of safety is planning and anticipating various possibilities and preparing for the eventuality. Any Physics teacher/educator planning on doing an experiment must try it before the students do it. In the event of an emergency, quick action can prevent many disasters. The teacher/educator must be willing to seek help at once without any hesitation because sometimes it may not be clear that the situation is hazardous and potentially dangerous.

There are a number of procedures to prevent and correct any hazardous situation. There are several safety aids available commercially such as posters, safety contracts, safety tests, safety citations, texts on safety in secondary classroom/laboratories, hand books on safety and a host of other equipment. Another important thing is to check the laboratory and classroom for safety and report it to the administrators before staring activities/experiments. We will discuss below areas that need special attention to safety.

1. Electricity: Safety in this area starts with locating the main cut off switch. All the power points, switches, and electrical connections must be checked one by one. Batteries and live wires must be checked. All checking must be done with the power turned off. The last act of assembling is to insert the plug and the first act of disassembling is to take off the plug.

2. Motion and forces: All stationary devices must be secured by C-clamps. Protective goggles must be used. Care must be taken at all times while knives, glass rods and heavy weights are used. Viewing a solar eclipse must always be indirect. When using model rockets, NASA's safety code must be implemented.

3. Heat: The master gas valve must be off at all times except while in use. Goggles and insulated gloves are to be used whenever needed. Never use closed containers for heating. Burners and gas connections must be checked periodically. Gas jets must be closed soon after the experiment is over. Fire retardant pads and quality glassware such as Pyrex must be used.

4. Pressure: While using a pressure cooker, never allow pressure to exceed 20 lb/square inch. The pressure cooker must be cooled before it is opened. Care must be taken when using mercury since it is poisonous. A drop of oil on mercury will prevent the mercury vapors from escaping.

5. Light: Broken mirrors or those with jagged edges must be discarded immediately. Sharp-edged mirrors must be taped. Spectroscopic light voltage connections must be checked periodically. Care must be taken while using ultraviolet light sources. Some students may have psychological or physiological reactions to the effects of strobe like (e.g. epilepsy).

6. Lasers: Direct exposure to lasers must not be permitted. The laser target must be made of non-reflecting material. The movement of students must be restricted during experiments with lasers. A number of precautions while using lasers must be taken – use of low power lasers, use of approved laser goggles, maintaining the room's brightness so that the pupils of the eyes remain small. Appropriate beam stops must be set up to terminate the laser beam when needed. Prisms should be set up before class to avoid unexpected reflection.

7. Sound: Fastening of the safety disc while using the high speed siren disc is very important. Teacher must be aware of the fact that sounds higher than 110 decibels will cause damage to hearing.

8. Radiation: Proper shielding must be used while doing experiments with x-rays. All tubes that are used in a physics laboratory such as vacuum tubes, heat effect tubes, magnetic or deflection tubes must be checked and used for demonstrations by the teacher. Cathode rays must be enclosed in a frame and only the teacher should move them from their storage space. Students must watch the demonstration from at least eight feet away.

9. Radioactivity: The teacher must be knowledgeable and properly trained to handle the equipment and to demonstrate. Proper shielding of radioactive material and proper handling of material are absolutely critical. Disposal of any radioactive material must comply with the guidelines of NRC.

It is important that teachers and educators follow these guidelines to protect the students and to avoid most of the hazards. They have a responsibility to protect themselves as well. **There should be not any compromises in issues of safety.**

Skill 18.2 Design appropriate laboratory investigations to study the principles and applications of physics.

The study of physics covers a broad range of areas, and the experiments used in the laboratory to help develop and validate theories must be designed in accordance with the particular subject being examined. Several broad concepts, however, are applicable, regardless of the particulars of the theory or experiment.

Disturbance by observation

When designing an experiment, it is crucial to take note of the impact that the very act of observation or measurement has upon the behavior of the objects under investigation. It may be said that an experiment does not involve purely a situation where a scientist observes some system, but, instead, the scientist is observing a system under the observation of a scientist. That is to say, the act of observation has an effect on the system being observed, and this must be recognized in analyzing and reporting results. In some cases, this does not have a significant impact, such as with many macroscopic measurements or experiments. Nevertheless, the effect can be profound in microscopic experiments, especially at the level of subatomic particles, where using light to observe a particle, for instance, actually disturbs the particle substantially. A macroscopic example of this principle is the measurement of voltage or current in an electrical circuit. Although most voltmeters and ammeters do not present a large load to the circuit, they do load the circuit to some degree. Consequently, this loading can have an effect on the measurements, albeit a small one in most cases. Thus, it is critical to note the inherent limitations present when designing an experiment to test some hypothesis or theory.

Isolation of parameters

In addition to the above difficulty, it must be noted that it is not possible to completely isolate a particular parameter for experimental testing. Thus, for example, although a mass can be theoretically treated as the sole entity in the universe, in reality there are innumerable other masses distributed throughout the universe, each of which has an effect (by gravitation, for example) on the mass of interest. An appropriate experiment is thus designed to attempt to minimize outside and unwanted influences, although this may, at times, be impossible. In many cases, however, it is possible to construct an experimental apparatus that sufficiently isolates the system or object under observation to the extent that adequate measurement accuracy is possible. For example, when performing calorimetric experiments, the use of thermally isolating materials can help to eliminate heat loss into the environment during collection of measurements. Another instance is the use of electromagnetic shielding, perhaps through the use of metal sheets or screens, when performing electrical or magnetic experiments.

Measurement precision

Another concern regarding experimental design is the ability of the measurement equipment to measure the desired parameters with sufficient accuracy and precision. If the parameter that is being measured is beyond the range of the equipment by way of being either too large, too small or having variations that are too fine, the experiment can yield no useful information. The measurement equipment may be more or less technical; it may simply involve the use of the human eye, which has its own host of problems in some cases, or it may involve sophisticated equipment such as radiation or thermal energy detectors, pressure sensors or other devices. Regardless of the particulars of the experiment, it is critical that the measurement equipment be adequate for the specific tasks of the investigation. On the other hand, the risks or costs associated with equipment that is overly accurate or precise in making measurements may also rule out higher-end equipment. Thus, it is important to avoid both overkill and inadequacy when selecting equipment for a particular experiment.

Financial burden

Financial concerns are no less important than physical or scientific concerns. An experiment is only as useful as the possibility of its implementation, and the cost of an experiment is a critical consideration in its appropriate design. High-energy physics, for example, is perennially hampered by this difficulty, as the costs of constructing particle accelerators and other associated equipment, not to mention the costs of maintenance and personnel can be staggeringly high. Thus, the ideal experimental design is one that can be implemented at a minimum of expense, or, at least, at an affordable level of expense, such as to allow for repeatable experimentation and proper maintenance. While theorizing may involve little more than pencil and paper, investigations in the laboratory can involve tremendous costs; an appropriate experimental design for the investigation must, therefore, take financial costs into account.

Skill 18.3   Demonstrate knowledge of the uses of basic equipment to illustrate physical principles and phenomena.

Oscilloscope: An oscilloscope is a piece of electrical test equipment that allows signal voltages to be viewed as two-dimensional graphs of electrical potential differences plotted as a function of time.

The oscilloscope functions by measuring the deflection of a beam of electrons traveling through a vacuum in a cathode ray tube. The deflection of the beam can be caused by a magnetic field outside the tube or by electrostatic energy created by plates inside the tube. The unknown voltage or potential energy difference can be determined by comparing the electron deflection it causes to the electron deflection caused by a known voltage.

Oscilloscopes can also determine if an electrical circuit is oscillating and at what frequency. They are particularly useful for troubleshooting malfunctioning equipment. You can see the "moving parts" of the circuit and tell if the signal is being distorted. With the aid of an oscilloscope you can also calculate the "noise" within a signal and see if the "noise" changes over time.

Inputs of the electrical signal are usually entered into the oscilloscope via a coaxial cable or probes. A variety of transducers can be used with an oscilloscope that enable it to measure other stimuli including sound, pressure, heat, and light.

Voltmeter/Ohmmeter/Ammeter: A common electrical meter, typically known as a multimeter, is capable of measuring voltage, resistance, and current. Many of these devices can also measure capacitance (farads), frequency (hertz), duty cycle (a percentage), temperature (degrees), conductance (siemens), and inductance (henrys).

These meters function by utilizing the following familiar equations:

Across a resistor (Resistor R):

$$V_R = IR_R$$

Across a capacitor (Capacitor C):

$$V_C = IX_C$$

Across an inductor (Inductor L):

$$V_L = IX_L$$

Where V=voltage, I=current, R=resistance, X=reactance.

If any two factors in the equations are held constant or are known, the third factor can be determined and is displayed by the multimeter.

Signal Generator: A signal generator, also known as a test signal generator, function generator, tone generator, arbitrary waveform generator, or frequency generator, is a device that generates repeating electronic signals in either the analog or digital domains. They are generally used in designing, testing, troubleshooting, and repairing electronic devices.

A function generator produces simple repetitive waveforms by utilizing a circuit called an electronic oscillator or a digital signal processor to synthesize a waveform. Common waveforms are sine, sawtooth, step or pulse, square, and triangular. Arbitrary waveform generators are also available which allow a user to create waveforms of any type within the frequency, accuracy and output limits of the generator. Function generators are typically used in simple electronics repair and design where they are used to stimulate a circuit under test. A device such as an oscilloscope is then used to measure the circuit's output.

Spectrometer: A spectrometer is an optical instrument used to measure properties of light over a portion of the electromagnetic spectrum. Light intensity is the variable that is most commonly measured but wavelength and polarization state can also be determined. A spectrometer is used in spectroscopy for producing spectral lines and measuring their wavelengths and intensities. Spectrometers are capable of operating over a wide range of wavelengths, from short wave gamma and X-rays into the far infrared. In optics, a spectrograph separates incoming light according to its wavelength and records the resulting spectrum in some detector. In astronomy, spectrographs are widely used with telescopes.

Skill 18.4   Use mathematical concepts, strategies, and procedures, including graphical and statistical methods and differential and integral calculus, to derive and manipulate formal relationships between physical quantities

Powers of ten	Order of magnitude
0.0001	−4
0.001	−3
0.01	−2
0.1	−1
1	0
10	1
100	2
1,000	3
10,000	4

The order of magnitude refers to a category of scale or size of an amount, where each category contains values of a fixed ratio to the categories before or after. The most common ratio is 10. The table to the left lists the orders of magnitude of the number 10 associated with the actual numbers.

Orders of magnitude are typically used to make estimations of a number. For example, if two numbers differ by one order of magnitude, one number is 10 times larger than the other. If they differ by two orders of magnitude the difference is 100 times larger or smaller, and so on. It follows that two numbers have the same order of magnitude if they differ by less than 10 times the size.

To estimate the order of manitude of a physical quantity, you round the its value to the nearest power of 10. For example, in estimating the human population of the earth, you may not know if it is 5 billion or 12 billion, but a reasonable order of magnitude estimate is 10 billion. Similarly, you may know that Saturn is much larger than Earth and can estiamte that it has approximatly 100 times more mass, or that its mass is 2 orders of magnitude larger. The actual number is 95 times the mass of earth.

Physical Item	Size	Order of Magnitude (meters)
Diameter of a hydrogen atom	100 picometers	10^{-10}
Size of a bacteria	1 micrometer	10^{-6}
Size of a raindrop	1 millimeter	10^{-3}
Width of a human finger	1 centimeter	10^{-2}
Height of Washinton Monument	100 meters	10^{2}
Height of Mount Everest	10 kilometers	10^{4}
Diameter of Earth	10 million meters	10^{7}
One light year	1 light year	10^{16}

The slope or the gradient of a line is used to describe the measurement of the steepness, incline, or grade. A higher slope value indicates a steeper incline. The slope is defined as the ratio of the "rise" divided by the "run" between two points on a line, that is to say, the ratio of the altitude change to the horizontal distance between any two points on the line.

The slope of a line in the plane containing the x and y axes is generally represented by the letter m, and is defined as the change in the y coordinate divided by the corresponding change in the x coordinate, between two distinct points on the line. This is described by the following equation:

$$m = \frac{\Delta y}{\Delta x}$$

Given two points (x1, y1) and (x2, y2), the change in x from one to the other is x2 - x1, while the change in y is y2 - y1. Substituting both quantities into the above equation obtains the following:

$$m = \frac{y_2 - y_1}{x_2 - x_1}$$

For example: if a line runs through two points: P(1,2) and Q(13,8). By dividing the difference in y-coordinates by the difference in x-coordinates, one can obtain the slope of the line:

$$m = \frac{\Delta y}{\Delta x} = \frac{y_2 - y_1}{x_2 - x_1} = \frac{8 - 2}{13 - 1} = \frac{6}{12} = \frac{1}{2}$$

The slope is 1/2 = 0.5.

The slope of curved lines can be approximated by selecting x and y values that are very close together. In a curved region the slope changes along the curve.

If we let Δx and Δy be the x and y distances between two points on a curve, then Δy /Δx is the slope of a secant line to the curve.

For example, the slope of the secant intersecting
$y = x^2$ at (0,0) and (3,9)

is m = (9 - 0) / (3 - 0) = 3

By moving the two points closer together so that Δy and Δx decrease, the secant line more closely approximates a tangent line to the curve, and as such the slope of the secant approaches that of the tangent.

In differential calculus, the derivative is essentially taking the change in y with respect to the change in x as the change in x approaches the limit of zero. The derivative of the curved line function is a line tangent to that point on the curve and is equal to the slope of the graph at that specific point.
In calculus, the integral of a function is an extension of the concept of summing and is given by the area under a graphical representation of the function. The integral is usually used to find a measure of totality such as area, volume, mass, or displacement when the rate of change is specified, as in any simple x-y graphical representation.

The simplest graph to analyze the area under is a flat horizontal line. As an example, let's say f is the constant function f(x) = 3 and we want to find the area under the graph from x= 0 to x=10. This is simply a rectangle 3 units high by 10 units long, or 30 units square. The same result can be found by integrating the function, though this is usually done for more complicated or smooth curves.

Let us imagine the curve of a function f(X) between X=0 and X=10. One way to approximate the area under the curve is to draw numerous rectangles under the curve of a given width, estimate their height and sum the area of each rectangle. We can say that the width of each rectangle is δX. But since the top of each column is not exactly straight, this is only an approximation.

When we use integral calculus to determine an integral, we are taking the limit of δX approaching zero, so there will be more and more columns which are thinner and thinner to fill the space between X=0 and X=10. The top of each column then gets closer and closer to being a straight line and our expression for the area therefore gets closer and closer to being exactly right.

Skill 18.5 Demonstrate an understanding of the growth of physics knowledge from a historical perspective.

Archimedes
Archimedes was a Greek mathematician, physicist, engineer, astronomer, and philosopher. He is credited with many inventions and discoveries some of which are still in use today such as the Archimedes screw. He designed the compound pulley, a system of pulleys used to lift heavy loads such as ships.

Although Archimedes did not invent the lever, he gave the first rigorous explanation of the principles involved which are the transmission of force through a fulcrum and moving the effort applied through a greater distance than the object to be moved. His Law of the Lever states that magnitudes are in equilibrium at distances reciprocally proportional to their weights.

He also laid down the laws of flotation and described Archimedes' principle which states that a body immersed in a fluid experiences a buoyant force equal to the weight of the displaced fluid.

Amedeo Avogadro
Avogadro was an Italian professor of physics born in the 18th century. He contributed to the understanding of the difference between atoms and molecules and the concept of molarity. The famous Avogadro's principle states that equal volumes of all gases at the same temperature and pressure contain an equal number of molecules.

Niels Bohr
Bohr was a Danish physicist who made fundamental contributions to understanding atomic structure and quantum mechanics. Bohr is widely considered one of the greatest physicists of the twentieth century.
Bohr's model of the atom was the first to place electrons in discrete quantized orbits around the nucleus.

Bohr also helped determine that the chemical properties of an element are largely determined by the number of electrons in the outer orbits of the atom. The idea that an electron could drop from a higher-energy orbit to a lower one emitting a photon of discrete energy originated with Bohr and became the basis for future quantum theory.

He also contributed significantly to the Copenhagen interpretation of quantum mechanics. He received the Nobel Prize for Physics for this work in 1922.

Robert Boyle

Robert Boyle was born in Ireland in 1627 and was one of the most prominent experimentalists of his time. He was the first scientist who kept accurate logs of his experiments and though an alchemist himself, gave birth to the science of chemistry as a separate rigorous discipline. He is well known for Boyle's law that describes the relationship between the pressure and volume of an ideal gas. It was one of the first mathematical expressions of a scientific principle.

Marie Curie

Curie was as a Polish-French physicist and chemist. She was a pioneer in radioactivity and the winner of two Nobel Prizes, one in Physics and the other in Chemistry. She was also the first woman to win the Nobel Prize.

Curie studied radioactive materials, particularly pitchblende, the ore from which uranium was extracted. The ore was more radioactive than the uranium extracted from it which led the Curies (Marie and her husband Pierre) to discover a substance far more radioactive then uranium. Over several years of laboratory work the Curies eventually isolated and identified two new radioactive chemical elements, polonium and radium. Curie refined the radium isolation process and continued intensive study of the nature of radioactivity.

Albert Einstein

Einstein was a German-born theoretical physicist who is widely considered one of the greatest physicists of all time. While best known for the theory of relativity, and specifically mass-energy equivalence, $E = mc^2$, he was awarded the 1921 Nobel Prize in Physics for his explanation of the photoelectric effect and "for his services to Theoretical Physics". In his paper on the photoelectric effect, Einstein extended Planck's hypothesis ($E = h\nu$) of discrete energy elements to his own hypothesis that electromagnetic energy is absorbed or emitted by matter in quanta and proposed a new law $E_{max} = h\nu - P$ to account for the photoelectric effect.

He was known for many scientific investigations including the special theory of relativity which stemmed from an attempt to reconcile the laws of mechanics with the laws of the electromagnetic field. His general theory of relativity considered all observers to be equivalent, not only those moving at a uniform speed. In general relativity, gravity is no longer a force, as it is in Newton's law of gravity, but is a consequence of the curvature of space-time.

Other areas of physics in which Einstein made significant contributions, achievements or breakthroughs include relativistic cosmology, capillary action, critical opalescence, classical problems of statistical mechanics and problems in which they were merged with quantum theory (leading to an explanation of the Brownian movement of molecules), atomic transition probabilities, the quantum theory of a monatomic gas, the concept of the photon, the theory of radiation (including stimulated emission), and the geometrization of physics.

Einstein's research efforts after developing the theory of general relativity consisted primarily of attempts to generalize his theory of gravitation in order to unify and simplify the fundamental laws of physics, particularly gravitation and electromagnetism, which he referred to as the Unified Field Theory.

Michael Faraday
Faraday was an English chemist and physicist who contributed significantly to the fields of electromagnetism and electrochemistry. He established that magnetism could affect rays of light and that the two phenomena were linked. It was largely due to his efforts that electricity became viable for use in technology. The unit for capacitance, the farad, is named after him as is the Faraday constant, the charge on a mole of electrons (about 96,485 coulombs). Faraday's law of induction states that a magnetic field changing in time creates a proportional electromotive force.

Sir Isaac Newton
Newton was an English physicist, mathematician, astronomer, alchemist, and natural philosopher in the late 17th and early 18th centuries. He described universal gravitation and the three laws of motion laying the groundwork for classical mechanics. He was the first to show that the motion of objects on earth and in space is governed by the same set of mechanical laws. These laws became central to the scientific revolution that took place during this period of history. Newton's three laws of motion are:

> I. Every object in a state of uniform motion tends to remain in that state of motion unless an external force is applied to it.
> II. The relationship between an object's mass m, its acceleration a, and the applied force F is $F = ma$.
> III. For every action there is an equal and opposite reaction.

In mechanics, Newton developed the basic principles of conservation of momentum. In optics, he invented the reflecting telescope and discovered that the spectrum of colors seen when white light passes through a prism is inherent in the white light and not added by the prism as previous scientists had claimed. Newton notably argued that light is composed of particles. He also formulated an experimental law of cooling, studied the speed of sound, and proposed a theory of the origin of stars.

J. Robert Oppenheimer

Oppenheimer was an American physicist, best known for his role as the scientific director of the Manhattan Project, the effort to develop the first nuclear weapons. Sometimes called "the father of the atomic bomb", Oppenheimer later lamented the use of atomic weapons. He became a chief advisor to the United States Atomic Energy Commission and lobbied for international control of atomic energy. Oppenheimer was one of the founders of the American school of theoretical physics at the University of California, Berkeley. He did important research in theoretical astrophysics, nuclear physics, spectroscopy, and quantum field theory.

Wilhelm Ostwald

Wilhelm Ostwald, born in 1853 in Latvia, was one of the founders of classical physical chemistry which deals with the properties and reactions of atoms, molecules and ions. He developed the Ostwald process for the synthesis of nitric acid. In 1909 he won the Nobel prize for his work on catalysis, chemical equilibria and reaction velocities.

Linus Pauling

The American chemist Linus Pauling won the Nobel prize for chemistry in 1954 for his investigation of the nature of the chemical bond. He led the way in applying quantum mechanics to chemistry. Later in his career he focused on biochemical problems such as the structure of proteins and sickle cell anemia. He won the Nobel peace prize in 1962 for his contribution to nuclear disarmament.

COMPETENCY 19.0 UNDERSTAND AND APPLY KNOWLEDGE OF PLANAR MOTION.

Skill 19.1 Analyze the relationship between vectors and physical quantities and perform a variety of vector algebra operations.

Vector space is a collection of objects that have **magnitude** and **direction**. They may have mathematical operations, such as addition, subtraction, and scaling, applied to them. Vectors are usually displayed in boldface or with an arrow above the letter. They are usually shown in graphs or other diagrams as arrows. The length of the arrow represents the magnitude of the vector while the direction in which the arrow points shows the vector direction.

To **add two vectors** graphically, the base of the second vector is drawn from the point of the first vector as shown below with vectors **A** and **B**. The sum of the vectors is drawn as a dashed line, from the base of the first vector to the tip of the second. As illustrated, the order in which the vectors are connected is not significant as the endpoint is the same graphically whether **A** connects to **B** or **B** connects to **A**. This principle is sometimes called the parallelogram rule.

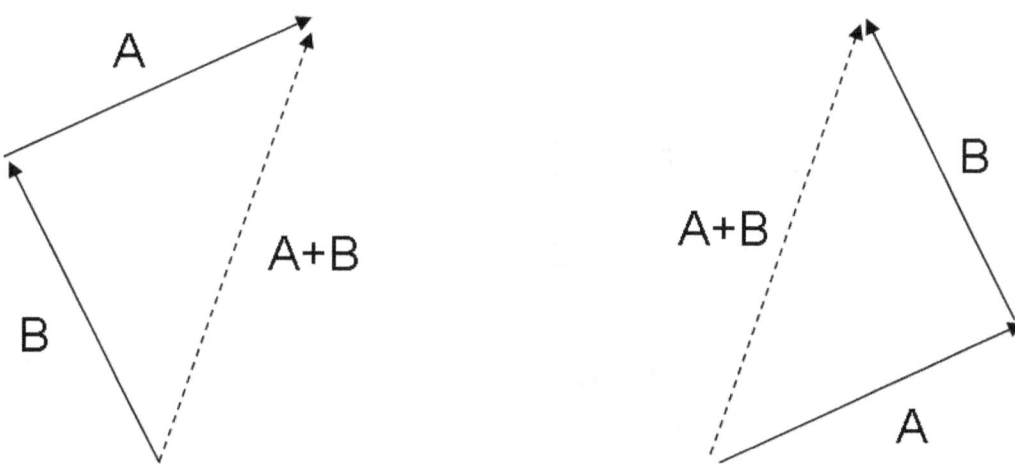

If more than two vectors are to be combined, additional vectors are simply drawn in accordingly with the sum vector connecting the base of the first to the tip of the final vector.

Subtraction of two vectors can be geometrically defined as follows. To subtract **A** from **B**, place the ends of **A** and **B** at the same point and then draw an arrow from the tip of **A** to the tip of **B**. That arrow represents the vector **B-A**, as illustrated below:

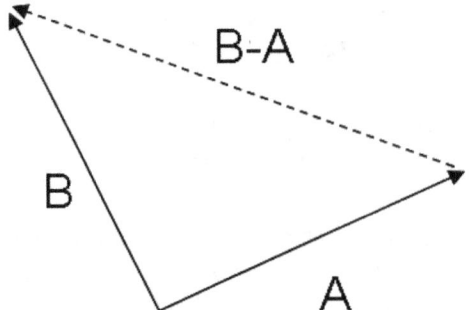

To add two vectors without drawing them, the vectors must be broken down into their orthogonal components using sine, cosine, and tangent functions. Add both x components to get the total x component of the sum vector, then add both y components to get the y component of the sum vector. Use the Pythagorean Theorem and the three trigonometric functions to the get the size and direction of the final vector.

Example: Here is a diagram showing the x and y-components of a vector D1:

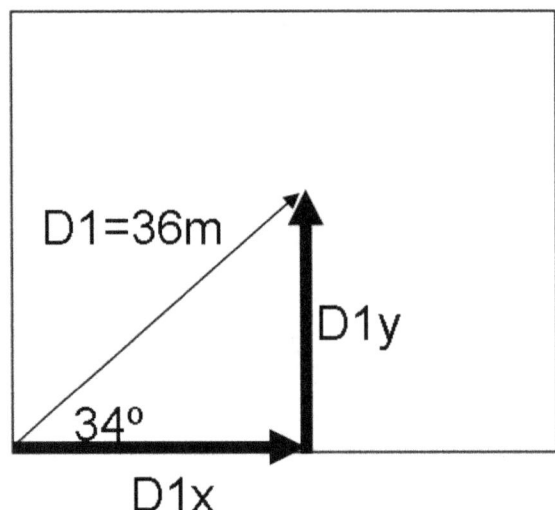

Notice that the x-component D1x is adjacent to the angle of 34 degrees.

Thus D1x=36m (cos34) =29.8m

The y-component is opposite to the angle of 34 degrees.

Thus D1y =36m (sin34) = 20.1m

A second vector D2 is broken up into its components in the diagram below using the same techniques. We find that D2y=9.0m and D2x=-18.5m.

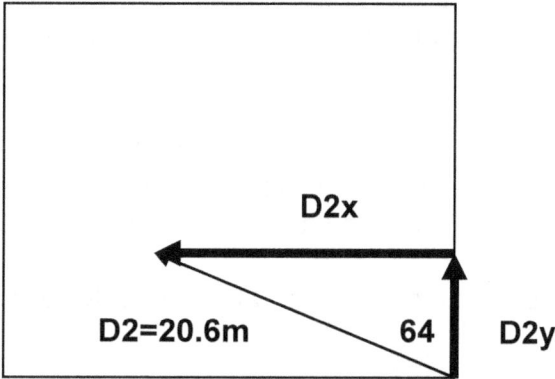

Next we add the x components and the y components to get

DTotal x = 11.3 m   and   DTotal y = 29.1 m

Now we have to use the Pythagorean theorem to get the total magnitude of the final vector. And the arctangent function to find the direction. As shown in the diagram below.

DTotal = 31.2m

tan θ = DTotal y / DTotal x = 29.1m / 11.3 = 2.6   θ = 69 degrees

Multiplication of Vectors

The **dot product** is also known as the scalar product. This is because the dot product of two vectors is not a vector, but a scalar (i.e., a real number without an associated direction). The definition of the dot product of the two vectors **a** and **b** is:

$$a \bullet b = \sum_{i=1}^{n} a_i b_i = a_1 b_1 + a_2 b_2 + ... + a_n b_n$$

The following is an example calculation of the dot product of two vectors:

[1 3 -5] · [4 -2 -2] = (1)(4) + (3)(-2) + (-5)(-2) = 8

Note that the product is a simple scalar quantity, not a vector. The dot product is commutative and distributive.

Unlike the dot product, the cross product does return another vector. The vector returned by the cross product is orthogonal to the two original vectors. The cross product is defined as:

$$\mathbf{a} \times \mathbf{b} = \mathbf{n}\, |\mathbf{a}|\, |\mathbf{b}|\, \sin \theta$$

where n is a unit vector perpendicular to both **a** and **b** and θ is the angle between **a** and **b**. In practice, the cross product can be calculated as explained below:

Given the orthogonal unit vectors **i**, **j**, and **k**, the vector **a** and **b** can be expressed:

$$\mathbf{a} = a_1 \mathbf{i} + a_2 \mathbf{j} + a_3 \mathbf{k}$$
$$\mathbf{b} = b_1 \mathbf{i} + b_2 \mathbf{j} + b_3 \mathbf{k}$$

Then we can calculate that

$$\mathbf{a} \times \mathbf{b} = \mathbf{i}(a_2 b_3) + \mathbf{j}(a_3 b_1) + \mathbf{k}(a_1 b_2) - \mathbf{i}(a_3 b_2) - \mathbf{j}(a_1 b_3) - \mathbf{k}(a_2 b_1)$$

The cross product is anticommutative (that is, **a** x **b** = - **b** x **a**) and distributive over addition.

Skill 19.2 Use algebra and calculus methods to determine the rectilinear displacement, velocity, and acceleration of particles and rigid bodies, given initial conditions.

Kinematics is the part of mechanics that seeks to understand the motion of objects, particularly the relationship between position, velocity, acceleration and time.

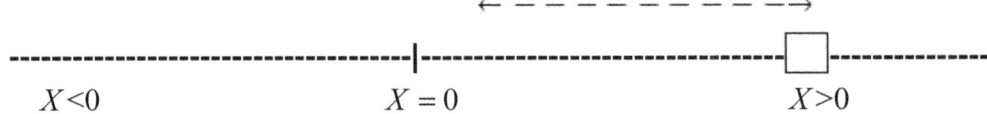

$X<0$ $X=0$ $X>0$

The above figure represents an object and its displacement along one linear dimension.

First we will define the relevant terms:

1. Position or Distance is usually represented by the variable x. It is measured relative to some fixed point or datum called the origin in linear units, meters, for example.

2. Displacement is defined as the change in position or distance which an object has moved and is represented by the variables D, d or Δx. Displacement is a vector with a magnitude and a direction.

3. Velocity is a vector quantity usually denoted with a V or v and defined as the rate of change of position. Typically units are distance/time, m/s for example. Since velocity is a vector, if an object changes the direction in which it is moving it changes its velocity even if the speed (the scalar quantity that is the magnitude of the velocity vector) remains unchanged.

i) Average velocity: $\vec{v} \equiv \dfrac{\Delta d}{\Delta t} = d_1 - d_0 / t_1 - t_0$

The ratio $\Delta d / \Delta t$ is called the average velocity. Average here denotes that this quantity is defined over a period Δt.

ii) Instantaneous velocity is the velocity of an object at a particular moment in time. Conceptually, this can be imagined as the extreme case when Δt is infinitely small.

5. Acceleration represented by a is defined as the rate of change of velocity and the units are m/s^2. Both an average and an instantaneous acceleration can be defined similarly to velocity.

From these definitions we develop the kinematic equations. In the following, subscript i denotes initial and subscript f denotes final values for a time period. Acceleration is assumed to be constant with time.

$$v_f = v_i + at \quad (1)$$

$$d = v_i t + \frac{1}{2}at^2 \quad (2)$$

$$v_f^2 = v_i^2 + 2ad \quad (3)$$

$$d = \left(\frac{v_i + v_f}{2}\right)t \quad (4)$$

Example:
Leaving a traffic light a man accelerates at 10 m/s². a) How fast is he going when he has gone 100 m? b) How fast is he going in 4 seconds? C) How far does he travel in 20 seconds.

Solution:
a) Use equation 3. He starts from a stop so v_i=0 and v_f^2=2 x 10m/s² x 100m=2000 m²/s² and v_f=45 m/s.
b) Use equation 1. Initial velocity is again zero so v_f=10m/s² x 4s=40 m/s.
c) Use equation 2. Since initial velocity is again zero, d=1/2 x 10 m/s² x (20s)²=2000 m

Skill 19.3 Use algebra and calculus methods to determine angular displacement, velocity, and acceleration of rigid bodies in a plane, given initial conditions.

Linear motion is measured in rectangular coordinates. Rotational motion is measured differently, in terms of the angle of displacement. There are three common ways to measure rotational displacement; degrees, revolutions, and radians. Degrees and revolutions have an easy to understand relationship, one revolution is 360°. Radians are slightly less well known and are defined as

$\frac{arc\ length}{radius}$. Therefore 360°=2π radians and 1 radian = 57.3°.

The major concepts of linear motion are duplicated in rotational motion with linear displacement replaced by **angular displacement**.

Angular velocity ω = rate of change of angular displacement.
Angular acceleration α = rate of change of angular velocity.

Also, the linear velocity v of a rolling object can be written as $v = r\omega$ and the linear acceleration as $a = r\alpha$.

One important difference in the equations relates to the use of mass in rotational systems. In rotational problems, not only is the mass of an object important but also its location. In order to include the spatial distribution of the mass of the object, a term called **moment of inertia** is used, $I = m_1 r_1^2 + m_2 r_2^2 + \cdots + m_n r_n^2$. The moment of inertia is always defined with respect to a particular axis of rotation.

Example:

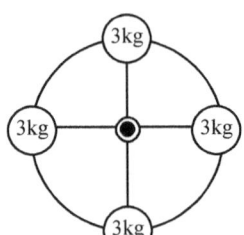

If the radius of the wheel on the left is 0.75m, what is its moment of inertia about an axis running through its center perpendicular to the plane of the wheel?

$$I = 3 \cdot 0.75^2 + 3 \cdot 0.75^2 + 3 \cdot 0.75^2 + 3 \cdot 0.75^2 = 6.75$$

Note: $I_{Sphere} = \frac{2}{5} mr^2$, $I_{Hoop/Ring} = mr^2$, $I_{disk} = \frac{1}{2} mr^2$

The rotational analog of Newton's second law of motion is given in terms of **torque** τ, moment of inertia I, and angular acceleration α:

$$\tau = I\alpha$$

where the torque τ is the rotational force on the body. In simple terms, the torque τ produced by a force F acting at a distance r from the point of rotation is given by the product of r and the component of the force that is perpendicular to the line joining the point of rotation to the point of action of the force.

A concept related to the moment of inertia is the **radius of gyration** (k), which is the average distance of the mass of an object from its axis of rotation, i.e., the distance from the axis where a point mass m would have the same moment of inertia.

$k_{Sphere} = \sqrt{\frac{2}{5}}r$, $k_{Hoop/Ring} = r$, $k_{disk} = \frac{r}{\sqrt{2}}$. As you can see $I = mk^2$

This is analogous to the concept of center of mass, the point where an equivalent mass of infinitely small size would be located, in the case of linear motion.

Angular momentum (L), and **rotational kinetic energy (KE$_r$)**, are therefore defined as follows: $L = I\omega$, $KE_r = \frac{1}{2}I\omega^2$

As with all systems, energy is conserved unless the system is acted on by an external force. This can be used to solve problems such as the one below.

Example:

A uniform ball of radius *r* and mass *m* starts from rest and rolls down a frictionless incline of height *h*. When the ball reaches the ground, how fast is it going?

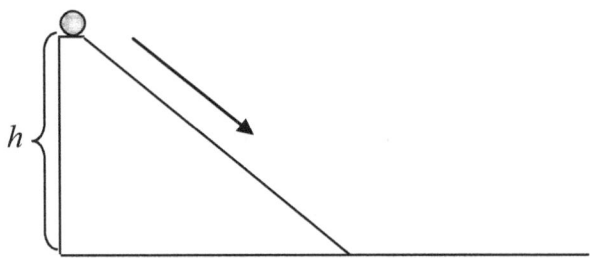

$$PE_{initial} + KE_{rotational/initial} + KE_{linear/initial} = PE_{final} + KE_{rotational/final} + KE_{linear/final}$$

$$mgh + 0 + 0 = 0 + \frac{1}{2}I\omega_{final}^2 + \frac{1}{2}mv_{final}^2 \rightarrow mgh = \frac{1}{2}\cdot\frac{2}{5}mr^2\omega_{final}^2 + \frac{1}{2}mv_{final}^2$$

$$mgh = \frac{1}{5}mr^2(\frac{v_{final}}{r})^2 + \frac{1}{2}mv_{final}^2 \rightarrow mgh = \frac{1}{5}mv_{final}^2 + \frac{1}{2}mv_{final}^2$$

$$gh = \frac{7}{10}v_{final}^2 \rightarrow v_{final} = \sqrt{\frac{10}{7}gh}$$

Similarly, unless a net torque acts on a system, the angular momentum remains constant in both magnitude and direction. This can be used to solve many different types of problems including ones involving satellite motion.

Example:
A planet of mass *m* is circling a star in an orbit like the one below. If its velocity at point A is 60,000m/s, and $r_B = 8\, r_A$, what is its velocity at point B?

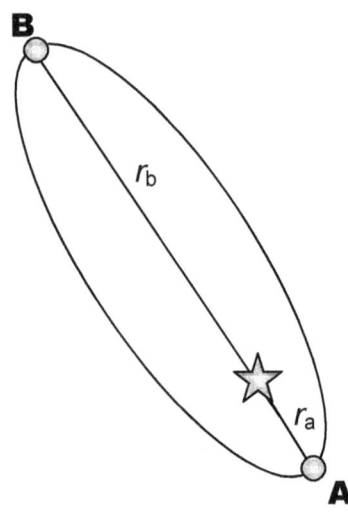

$$I_B\omega_B = I_A\omega_A$$
$$mr_B^2\omega_B = mr_A^2\omega_A$$
$$r_B^2\omega_B = r_A^2\omega_A$$
$$r_B^2\frac{v_B}{r_B} = r_A^2\frac{v_A}{r_A}$$
$$r_Bv_B = r_Av_A$$
$$8r_Av_B = r_Av_A$$
$$v_B = \frac{v_A}{8} = 7500m/s$$

Skill 19.4 Use algebra and calculus methods to determine the displacement, velocity, and acceleration of particles and rigid bodies undergoing periodic motion, given initial conditions.

Harmonic motion or harmonic oscillation is seen in any system that follows **Hooke's Law**. Hooke's law simply predicts the behavior of certain bodies as they return to equilibrium following a displacement. It is given by the following equation:

$$F = -kx$$

Where F=restoring force
x=displacement
k=a positive constant

From this equation we can see that the harmonic motion is neither damped nor driven. Harmonic motion is observed as sinusoidal oscillations about an equilibrium point. Both the amplitude and the frequency are constant. Further, the amplitude is always positive and is a function of the original force that disrupted the equilibrium. If an object's oscillation is governed solely by Hooke's law, it is a simple harmonic oscillator. Below are examples of simple harmonic oscillators:

Pendula: A pendulum is a mass on the end of a rigid rod or a string. An initial push will cause the pendulum to swing back and forth. This motion will be harmonic as long as the pendulum moves through an angle of less than 15°.

Masses connected to springs: A spring is simply the familiar helical coil of metal that is used to store mechanical energy. In a typical system, one end of a spring is attached to a mass and the other to a solid surface (a wall, ceiling, etc). If the spring is then stretched or compressed (i.e., removed from equilibrium) it will oscillate harmonically.

Vibrating strings: A string or rope tied tightly at both ends will oscillate harmonically when it is struck or plucked. This is often the mechanism used to generate sound in string-based instruments such as guitars and pianos.

Several simple harmonic oscillations maybe superimposed to create complex harmonic motion. The best-known example of complex harmonic motion is a musical chord.

View an animation of harmonic oscillation here:
http://en.wikipedia.org/wiki/Image:Simple_harmonic_motion_animation.gif

The displacement of a simple harmonic oscillator varies sinusoidally with time and is given by

$$x = A\cos(\omega t + \delta)$$

where A is the maximum displacement or amplitude, ω is the angular frequency and δ is the phase constant. We can see from the equation that the displacement goes through a full cycle at time intervals given by the period $T = 2\pi / \omega$. The figure below displays a graphical representation of the displacement of a simple harmonic oscillator.

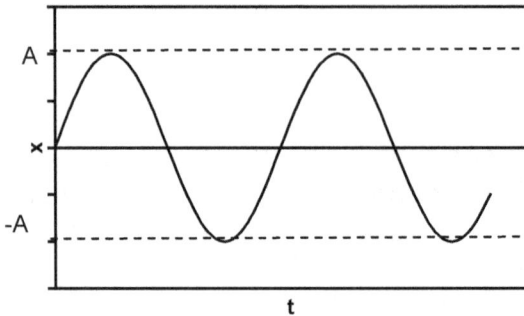

In the absence of dissipative forces, the total energy of a simple harmonic oscillator remains constant; however, the proportion of kinetic to potential energy varies. At x = A and x = -A all of the energy is potential. At x = 0 all of the energy is kinetic.

Skill 19.5 Analyze and solve problems involving the relationships of linear and angular displacement, velocity and acceleration.

Simple problems involving distance, displacement, speed, velocity, and constant acceleration can be solved by applying the kinematics equations from the proceeding section. The following steps should be employed to simplify a problem and apply the proper equations:

1. Create a simple diagram of the physical situation.
2. Ascribe a variable to each piece of information given.
3. List the unknown information in variable form.
4. Write down the relationships between variables in equation form.
5. Substitute known values into the equations and use algebra to solve for the unknowns.
6. Check your answer to ensure that it is reasonable.

Example:
A man in a truck is stopped at a traffic light. When the light turns green, he accelerates at a constant rate of 10 m/s². **a)** How fast is he going when he has gone 100 m? **b)** How fast is he going after 4 seconds? **c)** How far does he travel in 20 seconds?

a=10 m/s²

v_i=0 m/s

Solution:
We first construct a diagram of the situation.
In this example, the diagram is very simple, only showing the truck accelerating at the given rate. Next we define variables for the known quantities (these are noted in the diagram):

$$a=10 \text{ m/s}^2; \quad v_i=0 \text{ m/s}$$

Now we will analyze each part of the problem, continuing with the process outlined above.

For part **a)**, we have one additional known variable: d=100 m

The unknowns are: v_f (the velocity after the truck has traveled 100m)

Equation (3) will allow us to solve for v_f, using the known variables:

$$v_f^2 = v_i^2 + 2ad$$

$$v_f^2 = (0m/s)^2 + 2(10m/s^2)(100m) = 2000 \frac{m^2}{s^2}$$

$$v_f = 45 \frac{m}{s}$$

We use this same process to solve part **b)**. We have one additional known variable: t=4 s

The unknowns are: v_f (the velocity after the truck has traveled for 4 seconds)

Thus, we can use equation (1) to solve for v_f:

$$v_f = v_i + at$$

$$v_f = 0m/s + (10m/s^2)(4s) = 40m/s$$

For part **c)**, we have one additional known variable: t= 20 s

The unknowns are: d (the distance after the truck has traveled for 20 seconds)

Equation (2) will allow us to solve this problem:

$$d = v_i t + \frac{1}{2}at^2$$

$$d = (0m/s)(20s) + \frac{1}{2}(10m/s^2)(20s)^2 = 2000m$$

Finally, we consider whether these solutions seem physically reasonable. In this simple problem, we can easily say that they do.

Skill 19.6 Analyze and solve problems involving periodic motion and uniform circular motion

Motion on an arc can also be considered from the view point of the kinematic equations. As pointed out earlier, displacement, velocity and acceleration are all vector quantities, i.e. they have magnitude (the speed is the magnitude of the velocity vector) and direction. This means that if one drives in a circle at constant speed one still experiences an acceleration that changes the direction. We can define a couple of parameters for objects moving on circular paths and see how they relate to the kinematic equations.

1. Tangential speed: The tangent to a circle or arc is a line that intersects the arc at exactly one point. If you were driving in a circle and instantaneously moved the steering wheel back to straight, the line you would follow would be the tangent to the circle at the point where you moved the wheel. The tangential speed then is the instantaneous magnitude of the velocity vector as one moves around the circle.

2. Tangential acceleration: The tangential acceleration is the component of acceleration that would change the tangential speed and this can be treated as a linear acceleration if one imagines that the circular path is unrolled and made linear.

3. Centripetal acceleration: Centripetal acceleration corresponds to the constant change in the direction of the velocity vector necessary to maintain a circular path. Always acting toward the center of the circle, centripetal acceleration has a magnitude proportional to the tangential speed squared divided by the radius of the path.

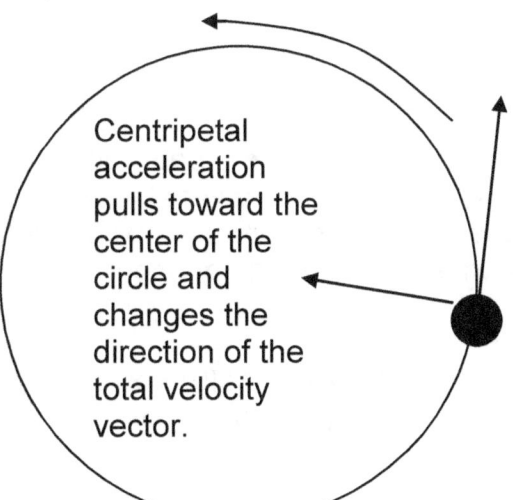

Centripetal acceleration pulls toward the center of the circle and changes the direction of the total velocity vector.

Tangential Speed= the magnitude of the velocity vector. A tangential acceleration changes the tangential speed.

TEACHER CERTIFICATION STUDY GUIDE

COMPETENCY 20.0   UNDERSTAND AND APPLY KNOWLEDGE OF FORCE, MOMENTUM, AND ENERGY AS THEY APPLY TO PLANAR MOTION.

Skill 20.1   Apply Newton's laws of motion to analyze and solve problems involving translational, rotational, and periodic motion.

Uniform circular motion describes the motion of an object as it moves in a circular path at constant speed. There are many everyday examples of this behavior though we may not recognize them if the object does not complete a full circle. For example, a car rounding a curve (that is an arc of a circle) often exhibits uniform circular motion.

The following diagram and variable definitions will help us to analyze uniform circular motion.

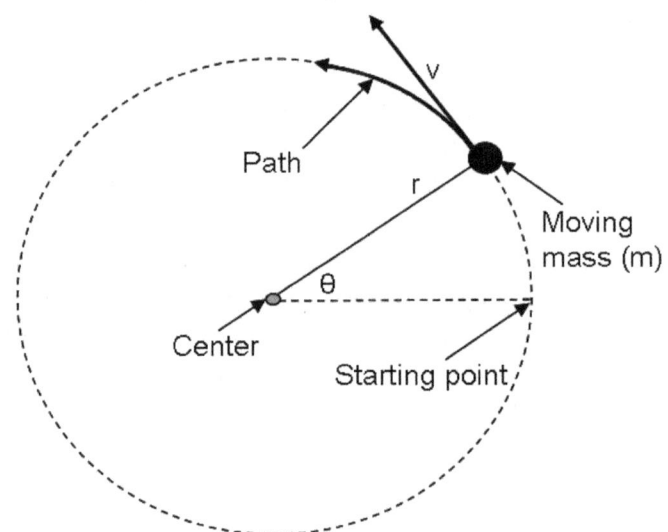

Above we see that the mass is traveling a path with constant radius (r) from some center point (x_0, y_0). By defining a variable (θ) that is a function of time (t) and is the angle between the mass's present position and original position on the circular path, we can write the following equations for the mass's position in a Cartesian plane.

$$x = r \cos(\theta) + x_0$$
$$y = r \sin(\theta) + y_0$$

Next observe that, because we are discussing uniform circular motion, the *magnitude* of the mass's velocity (v) is constant. However, the velocity's direction is always tangent to the circle and so always changing. We know that a changing velocity means that the mass must have a positive acceleration. This acceleration is directed toward the center of the circular path and is always perpendicular to the velocity, as shown below:

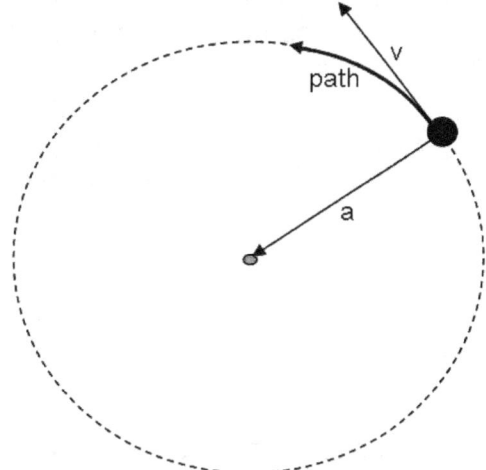

This is known as centripetal acceleration and is mathematically expressed as:

$$a = \frac{v^2}{r} = \frac{4\pi^2 r}{t^2}$$

where t is the period of the motion or the time taken for the mass to travel once around the circle. The force (F) experienced by the mass (m) is known as centripetal force and is always directed towards the center of the circular path. It has constant magnitude given by the following equation:

$$F = ma = m\frac{v^2}{r}$$

Skill 20.2 Apply the law of universal gravitation to solve problems involving free fall, projectile motion, and planetary motion.

Newton's universal law of gravitation states that any two objects experience a force between them as the result of their masses. Specifically, the force between two masses m_1 and m_2 can be summarized as

$$F = G\frac{m_1 m_2}{r^2}$$

where G is the gravitational constant ($G = 6.672 \times 10^{-11} Nm^2 / kg^2$), and r is the distance between the two objects.

Important things to remember:

1. The gravitational force is proportional to the masses of the two objects, but *inversely* proportional to the *square of the distance* between the two objects.
2. When calculating the effects of the acceleration due to gravity for an object above the earth's surface, the distance above the surface is ignored because it is inconsequential compared to the radius of the earth. The constant figure of 9.81 m/s² is used instead.

Problem: Two identical 4 kg balls are floating in space, 2 meters apart. What is the magnitude of the gravitational force they exert on each other?

Solution:

$$F = G\frac{m_1 m_2}{r^2} = G\frac{4 \times 4}{2^2} = 4G = 2.67 \times 10^{-10} \, N$$

For a satellite of mass m in orbit around the earth (mass M), the gravitational attraction of the earth provides the centripetal force that keeps the satellite in motion:

$$\frac{GMm}{r^2} = \frac{mv^2}{r} = mr\omega^2 = mr\left(\frac{2\pi}{T}\right)^2$$

Thus the period T of rotation of the satellite may be obtained from the equation

$$\frac{T^2}{r^3} = \frac{4\pi^2}{GM}$$

Skill 20.3 Analyze and solve problems involving the relationships between linear quantities and their rotational analogues.

Problems involving the relationships between linear and rotational quantities constitute a general class that also includes problems involving the relationships between the parameters of linear and angular motion. Generally speaking, analysis of these problems typically requires conversions between coordinate systems.

Rotational quantities, since they are often most easily expressed in terms of angles and radii rather than rectangular distances, are amenable to spherical or cylindrical coordinate systems. The unit vectors in these two coordinate systems are variable (with the exception of the $\hat{z}$ vector, which is a rectangular coordinate), and therefore change, depending on the location of interest.

Difficulties arise, for instance, when derivatives of quantities involving these unit vectors are considered. Since the derivatives of these variable unit vectors are not evident, it is usually necessary to convert them to rectangular coordinates, where the unit vectors are constant and can be taken outside the derivates. An example is shown below for the radial unit vector $\hat{r}$.

$$\frac{\partial \hat{r}}{\partial t} = \frac{\partial}{\partial t}[\hat{x}\cos\theta + \hat{y}\sin\theta] = \frac{\partial \theta}{\partial t}[-\hat{x}\sin\theta + \hat{y}\cos\theta] = \hat{\theta}\frac{\partial \theta}{\partial t}$$

The relationship of angular (rotational) motion and linear motion, specifically with regard to displacement, velocity and acceleration, are treated in Skill 19.5. Nevertheless, problems may involve more than just these quantities; momentum, kinetic energy, force (torque) and inertia may all be additional factors for consideration. Although the mathematics of rotational and linear quantities may differ, the principles of physics that apply to them are the same. In the case of the kinetic energy of an object, for instance, the rotational and linear quantities are both based on the same general principle. Most generally, the total kinetic energy of an object is determined by finding the kinetic energy of each infinitesimal portion of the object. In the linear case, each part moves with the same velocity, and thus the kinetic energy KE is given as follows, where m_{total} is the mass of the object and v is the velocity:

$$KE = \frac{1}{2}m_{total}v^2$$

In the case of a rotating object, different portions of the object move at different velocities. Therefore, a summation form (or, more generally, an integral) over all the separate portions i is required to calculate the rotational kinetic energy.

$$KE = \sum_i \frac{1}{2}m_i v_i^2$$

This is a general form (with the integral form being the most general) of the kinetic energy, which reduces to the linear form when the object undergoes only linear motion. A common rotational expression of this kinetic energy is given below, where I is the rotational inertia and ω the angular speed.

$$KE = \frac{1}{2}I\omega^2$$

Although one form is called "linear kinetic energy" and the other called "rotational kinetic energy," they are both energies of motion with no fundamental difference. As such, they must both, together with all other types of energy, be conserved.

Problems involving linear and rotational quantities are not necessarily fundamentally different, but are often treated differently, in a mathematical sense, for simplicity. Nevertheless, converting between linear and rotational quantities may be necessary during the course of solving or analyzing problems, as discussed above.

Skill 20.4 Solve problems involving the conservation of linear and angular momentum

The law of conservation of momentum states that the total momentum of an *isolated system* (not affected by external forces and not having internal dissipative forces) always remains the same. For instance, in any collision between two objects in an isolated system, the total momentum of the two objects after the collision will be the same as the total momentum of the two objects before the collision. In other words, any momentum lost by one of the objects is gained by the other. In one dimension this is easy to visualize.

Imagine two carts rolling towards each other as in the diagram below

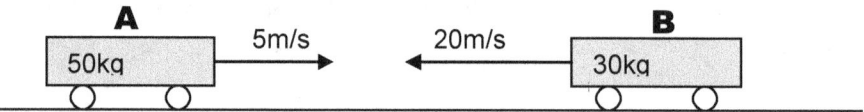

Before the collision, cart **A** has 250 kg m/s of momentum, and cart **B** has −600 kg m/s of momentum. In other words, the system has a total momentum of −350 kg m/s of momentum.

After the collision, the two cards stick to each other, and continue moving. How do we determine how fast, and in what direction, they go?

We know that the new mass of the cart is 80kg, and that the total momentum of the system is −350 kg m/s. Therefore, the velocity of the two carts stuck together must be $\frac{-350}{80} = -4.375 \, m/s$

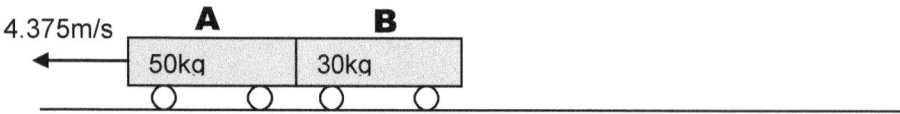

Conservation of momentum works the same way in two dimensions, the only change is that you need to use vector math to determine the total momentum and any changes, instead of simple addition.

PHYSICS

Imagine a pool table like the one below. Both balls are 0.5 kg in mass.

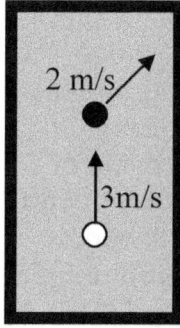

Before the collision, the white ball is moving with the velocity indicated by the solid line and the black ball is at rest.
After the collision the black ball is moving with the velocity indicated by the dashed line (a 135° angle from the direction of the white ball).

With what speed, and in what direction, is the white ball moving after the collision?

$p_{white/before} = .5 \cdot (0,3) = (0,1.5)$ $p_{black/before} = 0$ $p_{total/before} = (0,1.5)$

$p_{black/after} = .5 \cdot (2\cos 45, 2\sin 45) = (0.71, 0.71)$

$p_{white/after} = (-0.71, 0.79)$

i.e. the white ball has a velocity of $v = \sqrt{(-.71)^2 + (0.79)^2} = 1.06 m/s$

and is moving at an angle of $\theta = \tan^{-1}\left(\dfrac{0.79}{-0.71}\right) = -48°$ from the horizontal

See **Skill 19.3** for problems related to angular momentum.

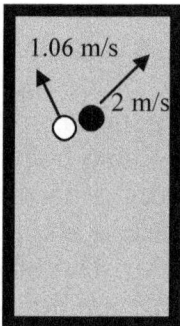

Skill 20.5 Use the relationship between work and energy, in algebraic and calculus forms, to solve problems involving the motions of physical systems acted upon by conservative and nonconservative forces.

The principle of conservation of energy states that an isolated system maintains a constant total amount of energy despite the fact that the energy may change forms. To put it another way, energy cannot be created or destroyed but can be changed from one form to another. For example, friction can turn kinetic energy into thermal energy. Other forms of energy include electrical energy, chemical energy, and mechanical energy.

A **conservative** force is one that conserves mechanical energy (kinetic + potential energy), i.e. there is no change in mechanical energy when a conservative force acts on an object. Consider a mass on a spring on a frictionless surface. This is a closed loop system. If conservative forces alone act on the mass during each cycle, the velocity of the mass at the beginning and the end of the cycle must be the same for the mechanical energy to have been conserved. In this way, the force has done no work. At any point in the cycle of motion, the total mechanical energy of the system remains constant even though the energy moves back and forth between kinetic and potential forms. If work is done on the mass, then the forces acting on the mass are **nonconservative**. In a real system there will be some dissipative forces that will convert some of the mechanical energy to thermal energy. Conservative forces are independent of path the object takes, while nonconservative forces are path dependent.

Gravity is a conservative force. This can be illustrated by imagining an object tossed into the air. On the upward journey the work done by gravity is the negative product of mass, acceleration, and height. On the downward journey, the work done by gravity is the positive value of this amount. Thus for the total loop the work is zero.

Friction is a nonconservative force. If a box is pushed along a rough surface from one side of the room to the other and back, friction opposes the movement in both directions; so the work done by friction cannot be equal to zero. This example also helps illustrate how nonconservative forces are path dependent. More work is done by friction if the path is tortuous rather than straight, even if the start and end points are the same. Let's try an example with a small box of mass 5 kg. The box moves in a circle 2 meters in diameter. The coefficient of kinetic friction between the box and the surface it rests on is 0.2. How much work is done by friction during one revolution?

The force exerted by friction is calculated by

$$F_k = \mu_k F_n = (0.2)(5 \text{ kg})(9.8 \text{ m/s}^2) = 9.8 \text{ N}$$

The force opposes the movement of the box during the entire distance of one revolution, or approximately 6.3 meters ($2\pi r$).
The total work done by friction is

$$W = F \times \cos\theta = (9.8 \text{ N})(6.3 \text{ m})(\cos 180) = -61.7 \text{ Joules}$$

As expected, the work is not zero since friction is not a conservative force. Since it does negative work on an object, it reduces the mechanical energy of the object and is a **dissipative** force.

COMPETENCY 21.0 UNDERSTAND AND APPLY KNOWLEDGE OF THE NATURE, PROPERTIES, AND BEHAVIOR OF MECHANICAL WAVES.

Skill 21.1 Apply the relationships among wave speed, wavelength, period, and frequency to analyze and solve problems related to wave propagation.

To fully understand waves, it is important to understand many of the terms used to characterize them.

Wave velocity: Two velocities are used to describe waves. The first is phase velocity, which is the rate at which a wave propagates. For instance, if you followed a single crest of a wave, it would appear to move at the phase velocity. The second type of velocity is known as group velocity and is the speed at which variations in the wave's amplitude shape propagate through space. Group velocity is often conceptualized as the velocity at which energy is transmitted by a wave. Phase velocity is denoted v_p and group velocity is denoted v_g. In a medium with refractive index independent of frequency, such as vacuum, the phase velocity is equal to the group velocity.

Crest: The maximum value that a wave assumes; the highest point.

Trough: The lowest value that a wave assumes; the lowest point.

Nodes: The points on a wave with minimal amplitude.

Antinodes: The farthest point from the node on the amplitude axis; both the crests and the troughs are antinodes.

Amplitude: The distance from the wave's highest point (the crest) to the equilibrium point. This is a measure of the maximum disturbance caused by the wave and is typically denoted by A.

Wavelength: The distance between any two sequential troughs or crests denoted λ and representing a complete cycle in the repeated wave pattern.

Period: The time required for a complete wavelength or cycle to pass a given point. The period of a wave is usually denoted T.

Frequency: The number of periods or cycles per unit time (usually a second). The frequency is denoted f and is the inverse of the wave's period (that is, $f=1/T$).

Phase: This is a given position in the cycle of the wave. It is most commonly used in discussing a "being out of phase" or a "phase shift", an offset between waves.

PHYSICS

We can visualize several of these terms on the following diagram of a simple, periodic sine wave on a scale of distance displacement (x-axis) vs. (y-axis):

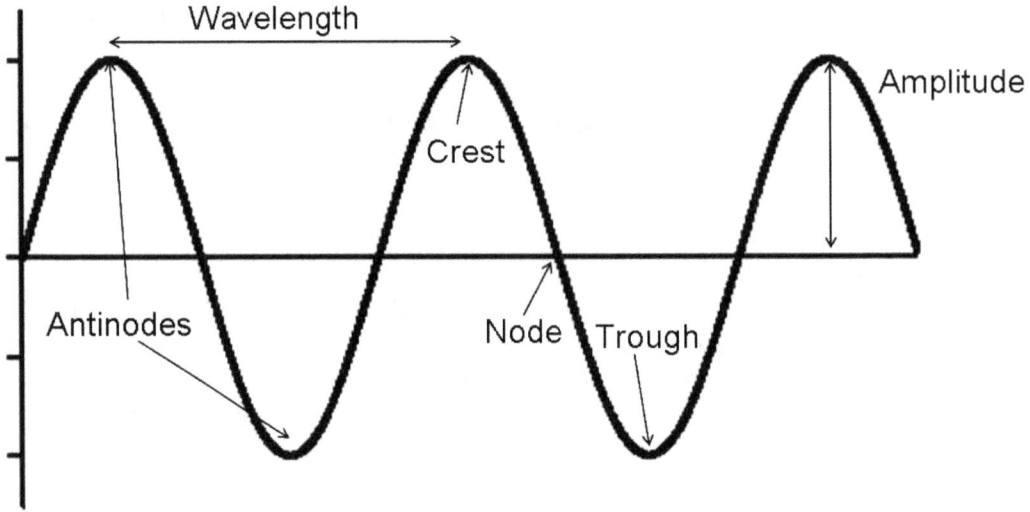

The phase velocity of a wave is related to its wavelength and frequency. Taking light waves, for instance, the speed of light c is equal to the distance traveled divided by time taken. Since the light wave travels the distance of one wavelength λ in the period of the wave T,

$$c = \frac{\lambda}{T}$$

The frequency of a wave, f, is the number of completed periods in one second. In general,

$$f = \frac{1}{T}$$

So the formula for the speed of light can be rewritten as

$$c = \lambda f$$

Thus the phase velocity of a wave is equal to the wavelength times the frequency.

Skill 21.2 Analyze the interference and reflection of waves and wave pulses.

Interference occurs when two or more waves are superimposed. Usually, interference is observed in coherent waves, well-correlated waves that have very similar frequencies or even come from the same source.

Superposition of waves may result in either constructive or destructive interference. Constructive interference occurs when the crests of the two waves meet at the same point in time. Conversely, destructive interference occurs when the crest of one wave and the trough of the other meet at the same point in time. It follows, then, that constructive interference increases amplitude and destructive interference decreased it. We can also consider interference in terms of wave phase; waves that are out of phase with one another will interfere destructively while waves that are in phase with one another will interfere constructively. In the case of two simple sine waves with identical amplitudes, for instance, amplitude will double if the waves are exactly in phase and drop to zero if the waves are exactly 180° out of phase.

Additionally, interference can create a standing wave, a wave in which certain points always have amplitude of zero. Thus, the wave remains in a constant position. Standing waves typically results when two waves of the same frequency traveling in opposite directions through a single medium are superposed. View an animation of how interference can create a standing wave at the following URL:

http://www.glenbrook.k12.il.us/GBSSCI/PHYS/mmedia/waves/swf.html

All wavelengths in the EM spectrum can experience interference but it is easy to comprehend instances of interference in the spectrum of visible light. One classic example of this is Thomas Young's double-slit experiment. In this experiment a beam of light is shone through a paper with two slits and a striated wave pattern results on the screen. The light and dark bands correspond to the areas in which the light from the two slits has constructively (bright band) and destructively (dark band) interfered.

Similarly, we may be familiar with examples of interference in sound waves. When two sounds waves with slightly different frequencies interfere with each other, beat results. We hear a beat as a periodic variation in volume with a rate that depends on the difference between the two frequencies. You may have observed this phenomenon when listening to two instruments being tuned to match; beating will be heard as the two instruments approach the same note and disappear when they are perfectly in tune.

Skill 21.3 Describe and analyze the nature, production and transmission of sound waves in various uniform media.

Sound is a compression wave that requires a medium through which to propagate. Compressions occur in places in which the medium's particles move closer together, the rarefactions occur where the medium's particles spread apart.

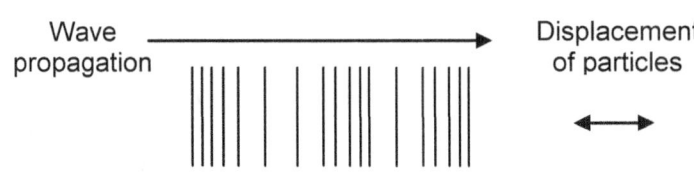

Sound in a Gas
Sound travels slowest through a gas. This is due to the fact that the particles in a gas are far apart, so the compression is not as significant. Sound also moves faster in warm air because the particles are more rapid in their movement. The speed of sound in a gas is a function of temperature. The speed of sound in dry air can be determined by the equation

$$v = 331.4 + 0.6T_c$$

where v= velocity of the wave in m/s
T_c is the temperature in °C

Sound in a liquid
Sound travels faster in a liquid than a gas because the particles in a liquid are more tightly packed, and therefore can transmit the compression wave more readily. On average, sound travels 4 times faster in a liquid than in a gas. Sound travels 1482 m/s in freshwater.

Sound in a solid
Sound travels fastest in solids as the rigid particle structure allows the compression waves to be transmitted very easily. The more dense the solid, the faster the wave will travel through the solid. There are a few solids that are less dense than liquids. In these solids, sound will not travel as fast as through the liquid. Sound travels 17 times faster in steel than in a gas.

The frequency of the harmonic acoustic wave, ω, determines the pitch of the sound in the same manner that the frequency determines the color of an electromagnetic harmonic wave. A high pitch sound corresponds to a high frequency sound wave and a low pitch sound corresponds to a low frequency sound wave. Similarly, the amplitude of the pressure determines the loudness or just as the amplitude of the electric (or magnetic) field E determines the brightness or intensity of a color. The intensity of a sound wave is proportional to the square of its amplitude.

The decibel scale is used to measure sound intensity. It originated in a unit known as the bel which is defined as the reduction in audio level over 1 mile of a telephone cable. Since the bel describes such a large variation in sound, it became more common to use the decibel, which is equal to 0.1 bel. A decibel value is related to the intensity of a sound by the following equation:

$$X_{dB} = 10 \log_{10}\left(\frac{X}{X_0}\right)$$

Where X_{dB} is the value of the sound in decibels
    X is the intensity of the sound
    X_0 is a reference value with the same units as X. X_0 is commonly taken to be the threshold of hearing at $10^{-12} W/m^2$.

It is important to note the logarithmic nature of the decibel scale and what this means for the relative intensity of sounds. The perception of the intensity of sound increases logarithmically, not linearly. Thus, an increase of 10 dB corresponds to an increase by one order of magnitude. For example, a sound that is 20 dB is not twice as loud as sound that is 10 dB; rather, it is 10 times as loud. A sound that is 30 dB will the 100 times as loud as the 10 dB sound.

Finally, let's equate the decibel scale with some familiar noises. Below are the decibel values of some common sounds.

Whispering voice: 20 dB
Quiet office: 60 dB
Traffic: 70 dB
Cheering football stadium: 110 dB
Jet engine (100 feet away): 150 dB
Space shuttle liftoff (100 feet away): 190 dB

Skill 21.4 Describe how the perception of sound depends on the physical properties of sound waves.

Pitch: The frequency of a sound wave as perceived by the human ear. A high pitch sound corresponds to a high frequency sound wave and a low pitch sound corresponds to a low frequency sound wave.

Sound power: This is the sonic energy of a sound wave per unit time.

Sound intensity: This is simply the sound power per unit area or the loudness of the sound.

A **standing wave** typically results from the interference between two waves of the same frequency traveling in opposite directions. The result is a stationary vibration pattern. One of the key characteristics of standing waves is that there are points in the medium where no movement occurs. The points are called nodes and the points where motion is maximal are called antinodes. This property allows for the analysis of various typical standing waves.

Vibrating string
Imagine a string of length L tied tightly at its two ends. We can generate a standing wave by plucking the string. The waves traveling along the string are reflected at the fixed end points and interfere with each other to produce standing waves. There will always be two nodes at the ends where the string is tied. Depending on the frequency of the wave that is generated, there may also be other nodes along the length of the string. In the diagrams below, examples are given of strings with 0, 1, or 2 additional nodes. These vibrations are known as the 1st, 2nd, and 3rd **harmonics**. The higher order harmonics follow the same pattern although they are not diagramed here.

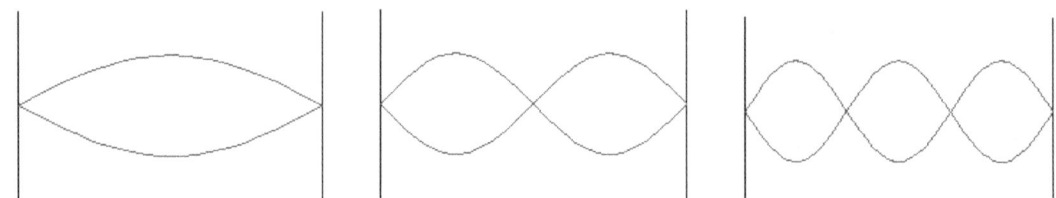

Since we know the length of the string (L) in each case, we can calculate the wavelength (λ) and frequency (f) for any harmonic using the following formula, where n=the harmonic order (n=1,2,3...) and v is the phase velocity of the wave.

$$\lambda_n = \frac{2L}{n} \qquad f_n = \frac{v}{\lambda_n} = n\frac{v}{2L}$$

Waves in a tube

Just as on a string, standing waves can propagate in gaseous or liquid medium inside a tube. In these cases we will also observe harmonic vibrations, but their nature will depend on whether the ends of the tube are closed or open. Specifically, an antinode will be observed at an open end and a node will appear at a closed end. Then, just as in the string example above, we can derive formulas that allow us to predict the wavelength and frequency of the harmonic vibrations that occur in a tube. Below, only frequencies are given; wavelength can be found by applying the formula f=v/λ.

For a tube with two closed ends or two open ends (note that this is the same as for the string described above):

$$f_n = \frac{nv}{2L}$$

where n = 1,2,3,4...

When only one end of a tube is closed, that end become a node and wave exhibits odd harmonics. For a tube with one close end and one open end:

$$f_n = \frac{nv}{4L}$$

where n = 1,3,5,7...

An animation at the following URL may be helpful in visualizing these various standing waves:

http://www.physics.smu.edu/~olness/www/05fall1320/applet/pipe-waves.html

When two sound waves with slightly different frequencies interfere with each other, **beats** result. We hear a beat as a periodic variation in volume with a rate that depends on the difference between the two frequencies. You may have observed this phenomenon when listening to two instruments being tuned to match; beating will be heard as the two instruments approach the same note and disappear when they are perfectly in tune.

TEACHER CERTIFICATION STUDY GUIDE

COMPETENCY 22.0   UNDERSTAND AND APPLY KNOWLEDGE OF THE NATURE, PROPERTIES, AND BEHAVIOR OF ELECTROMAGNETIC RADIATION.

Skill 22.1   Classify the regions of the electromagnetic spectrum relative to their frequency or wavelength.

The electromagnetic spectrum is measured using frequency (f) in hertz or wavelength (λ) in meters. The frequency times the wavelength of every electromagnetic wave equals the speed of light (3.0×10^8 meters/second).

Roughly, the range of wavelengths of the electromagnetic spectrum is:

	f	λ
Radio waves	$10^{-5} - 10^{-1}$ hertz	$10^3 - 10^9$ meters
Microwaves	$3 \times 10^9 - 3 \times 10^{11}$ hertz	$10^{-3} - 10^{-1}$ meters
Infrared radiation	$3 \times 10^{11} - 4 \times 10^{14}$ hertz	$7 \times 10^{-7} - 10^{-3}$ meters
Visible light	$4 \times 10^{14} - 7.5 \times 10^{14}$ hertz	$4 \times 10^{-7} - 7 \times 10^{-7}$ meters
Ultraviolet radiation	$7.5 \times 10^{14} - 3 \times 10^{16}$ hertz	$10^{-8} - 4 \times 10^{-7}$ meters
X-Rays	$3 \times 10^{16} - 3 \times 10^{19}$ hertz	$10^{-11} - 10^{-8}$ meters
Gamma Rays	$>3 \times 10^{19}$ hertz	$<10^{-11}$ meters

Radio waves are used for transmitting data. Common examples are television, cell phones, and wireless computer networks. Microwaves are used to heat food and deliver Wi-Fi service. Infrared waves are utilized in night vision goggles. Visible light we are all familiar with as the human eye is most sensitive to this wavelength range. Light of different colors have different wavelengths. In the visible range, red light has the largest wavelength while violet light has the smallest. UV light causes sunburns and would be even more harmful if most of it were not captured in the Earth's ozone layer. X-rays aid us in the medical field and gamma rays are most useful in the field of astronomy.

Skill 22.2   Analyze and predict the behavior of various types of electromagnetic radiation as they interact with matter.

Electromagnetic radiation is defined as the energy propagated through space between electric and magnetic fields. The electromagnetic spectrum is the extent of that energy ranging from cosmic rays, gamma rays, x-rays to ultraviolet, visible and infrared radiation including microwave energy. Electromagnetic waves are produced by the motion of electrically charged particles. These waves are also called "electromagnetic radiation" because they radiate from the electrically charged particles. They travel through empty space as well as through air and other substances. Scientists have observed that electromagnetic radiation acts like waves and also like a stream of particles (called "photons") that have no mass. The photons with the highest energy correspond to the shortest wavelengths. As the wavelength gets higher, the energy decreases.

PHYSICS

Wavelength and energy are inversely proportional. The full range of wavelengths (and photon energies) is called the "electromagnetic spectrum" (see **Skill 22.1**).

All electromagnetic radiation is a form of energy and is potentially destructive to matter. The energy can be calculated using the Planck relationship:

$$E = h \cdot f$$

> Where E= energy
> f=frequency
> h=Planck's constant (6.63×10^{-34} J·s)

From the above equation, it is clear that E and wavelength are inversely proportional (since frequency is inversely proportional to wavelength). Radio waves are of low energy and cosmic rays are of very high energy. All radiation interacts with matter and the study of the interaction of electromagnetic radiation with matter is known as spectroscopy.

Now let us take a look at the way radiation behaves and interacts with matter. Electromagnetic radiation can be broadly divided into two types: ionizing, which is capable of ionizing molecules and atoms, and non-ionizing, which cannot. Examples of each type are provided below:

Ionizing

1. Radio waves: These are capable of causing a change in the magnetic orientation of some atoms. This is known as Nuclear Magnetic Resonance (NMR) and is the basis of new medical application called magnetic Resonance Imaging (MRI). NMR spectroscopy is used in finding the structures of molecules. Atomic nuclei, e.g. hydrogen nuclei or proton (hydrogen is the only element which has no neutrons and in the nucleus of a hydrogen atom, there is only one proton) behave like little magnets and are capable of aligning themselves with or against an applied magnetic field. Because radio waves have very small mounts of energy, they are least hazardous. Radio waves from radio transmitters are passing through our body all the time.

2. Microwaves: These induce molecular tumbling and cause heating and temperature is a measure of the movement of molecules, more energy the molecules have, more movement is observed. When microwaves absorb matter, matter gets heat and becomes hot. It is very essential that microwave oven is leak proof, so that stray microwave radiation is not emitted from the oven or else it could cause localized heating of body fluids in a person near it.

3. Infrared radiation: This is of two types – near (shorter wavelengths) and far (longer wavelengths) infrared. Infrared radiation is used in meteorological satellite imagery. The most important thing as to why infrared radiation is used in imagery is that it temperature differences, which are vital meteorological studies. Higher temperatures indicate more infrared radiation and lower temperatures The absorption of infrared radiation by certain gases in the atmosphere gives rise to the "Greenhouse effect", which warms the atmosphere naturally. Without this, the temperature on earth would be below freezing point and all water would exist as ice. Infrared radiation causes molecular vibration, causing a rise in the temperature. Infrared spectroscopy is used in the study o molecular size and structure. The effect of infrared radiation on matter is heating, and if matter is exposed longer, results in burning just like toast in a toaster, left for a longer duration, gets burnt. The same thing happens to flesh as well.

4. Visible light: 41% of sunlight is considered to be harmless, but still contains energy. Visible reacts with matter in a way that causes electronic excitation, leading to color. Visible light has not got enough energy to ionize matter it comes in contact with. The most important reaction on earth, Photosynthesis is due to the absorption of visible light by plants and the excitation of molecules, in turn leading to the synthesis of sugars.

Ionizing Radiation

1. Ultraviolet radiation: UV radiation is the lowest frequency EM radiation that can trigger ionizing. UV is sometimes sub-divided into UV-A (315-400nm), UV-B (280-315nm) and UV -C (<280nm). UV-C is the most energetic and dangerous and is almost completely absorbed by the ozone layer in the stratosphere as is some of the UV -B. However, the thinning of ozone layer is resulting in increased amounts of UV -B reaching the earth's surface, causing increased erythema (sun burn), snow blindness, cataracts and skin cancer. UV -radiation cause electrons in the matter to excite, resulting in rupture of chemical bonds and thus forms unpaired electrons known as free radicals, which are highly reactive and damage DNA and thereby cell division. UV light is capable of breaking chemical bonds and brings about many chemical reactions, which are called "Photochemical reactions" – e.g. the bleaching of colored fabric in sunlight and yellowing of cotton etc. "Photochemical smog" is another example of photochemical reactions.

2. X-rays: These are very penetrating, of short wavelength, high energy, electromagnetic radiation that readily pass through soft tissues of humans because lighter atoms like hydrogen, carbon, nitrogen and oxygen atoms are transparent to x-rays. Atoms with heavier nuclei are less transparent and this makes x-rays to travel less well through bony tissues containing calcium and phosphorous.

3. Gamma-rays (γ): These have the highest frequency and energy, and also the shortest wavelength in the electromagnetic radiation spectrum, which are high energy photons. Because of their high energy content, they are capable of causing serious damage when they come into contact with living matter, living things.

Ionizing radiation can be extremely damaging to biological tissues. Particularly it is harmful to DNA, which may be damaged to an extent to cause cell death. Non-lethal DNA damage can trigger unrestricted cell division leading to cancer. However, ionizing radiation is used in technology such as the radiation therapy that kill cancer cells, radiography that detects internal defects of a materials, tracing materials in nuclear medicine, and sterilizing food and medical hardware.

Skill 22.3 Analyze and predict the behaviors of light, including interference, reflection, diffraction, polarization, and refraction.

According to the **principle of linear superposition**, when two or more waves exist in the same place, the resultant wave is the sum of all the waves, i.e. the amplitude of the resulting wave at a point in space is the sum of the amplitudes of each of the component waves at that point.

Interference occurs when two or more waves are superimposed. Usually, interference is observed in coherent waves, well-correlated waves that have very similar frequencies or even come from the same source. Superposition of waves may result in either constructive or destructive interference. Constructive interference occurs when the crests of the two waves meet at the same point in time. Conversely, destructive interference occurs when the crest of one wave and the trough of the other meet at the same point in time. It follows, then, that constructive interference increases amplitude and destructive interference decreases it. We can also consider interference in terms of wave phase; waves that are out of phase with one another will interfere destructively while waves that are in phase with one another will interfere constructively. In the case of two simple sine waves with identical amplitudes, for instance, amplitude will double if the waves are exactly in phase and drop to zero if the waves are exactly 180° out of phase.

Additionally, interference can create a **standing wave**, a wave in which certain points always have amplitude of zero. Thus, the wave remains in a constant position. Standing waves typically results when two waves of the same frequency traveling in opposite directions through a single medium are superposed. View an animation of how interference can create a standing wave at the following URL:

http://www.glenbrook.k12.il.us/GBSSCI/PHYS/mmedia/waves/swf.html

All wavelengths in the EM spectrum can experience interference but it is easy to comprehend instances of interference in the spectrum of visible light. One classic example of this is **Thomas Young's double-slit experiment**. In this experiment a beam of light is shone through a paper with two slits and a striated wave pattern results on the screen. The light and dark bands correspond to the areas in which the light from the two slits has constructively (bright band) and destructively (dark band) interfered. Light from any source can be used to obtain interference patterns. For example, Newton's rings can be produced with sun light. However, in general, white light is less suited for producing clear interference patterns as it is a mix of a full spectrum of colors. Sodium light is close to monochromatic and is thus more suitable for producing interference patterns. The most suitable is laser light as it is almost perfectly monochromatic.

Problem: The interference maxima (location of bright spots created by constructive interference) for double-slit interference are given by

$$\frac{n\lambda}{d} = \frac{x}{D} = \sin\theta \quad n=1,2,3...$$

where λ is the wavelength of the light, d is the distance between the two slits, D is the distance between the slits and the screen on which the pattern is observed and x is the location of the nth maximum. If the two slits are 0.1mm apart, the screen is 5m away from the slits, and the first maximum beyond the center one is 2.0 cm from the center of the screen, what is the wavelength of the light?

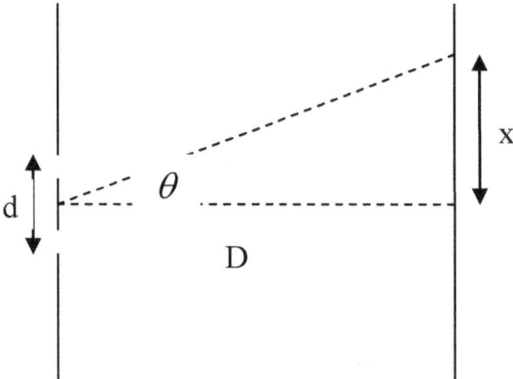

Solution: λ = xd/(Dn) = 0.02 x 0.0001/ (5 x1) = 400 nanometers

Thin-film interference occurs when light waves reflecting off the top surface of a film interfere with the waves reflecting from the bottom surface. We see colors in soap bubbles or in oil films floating on water since the criteria for constructive or destructive interference depend on the wavelength of the light. Non-reflective coatings on materials make use of destructive thin-film interference.

Diffraction is an important characteristic of waves. This occurs when part of a wave front is obstructed. Diffraction and interference are essentially the same physical process. Diffraction refers to various phenomena associated with wave propagation such as the bending, spreading, and interference of waves emerging from an aperture. It occurs with any type of wave including sound waves, water waves, and electromagnetic waves such as light and radio waves.

Here, we take a close look at important phenomena like single-slit diffraction, double-slit diffraction, diffraction grating, other forms of diffraction and lastly interference.

1. Single-slit diffraction: The simplest example of diffraction is single-slit diffraction in which the slit is narrow and a pattern of semi-circular ripples is formed after the wave passes through the slit.

2. Double-slit diffraction: These patterns are formed by the interference of light diffracting through two narrow slits.

3. Diffraction grating: Diffraction grating is a reflecting or transparent element whose optical properties are periodically modulated. In simple terms, diffraction gratings are fine parallel and equally spaced grooves or rulings on a material surface. When light is incident on a diffraction grating, light is reflected or transmitted in discrete directions, called diffraction orders. Because of their light dispersive properties, gratings are commonly used in monochromators and spectrophotometers. Gratings are usually designated by their groove density, expressed in grooves/millimeter. A fundamental property of gratings is that the angle of deviation of all but one of the diffracted beams depends on the wavelength of the incident light.

4. Other forms of diffraction:
i) Particle diffraction: It is the diffraction of particles such as electrons, which is used as a powerful argument for quantum theory. It is possible to observe the diffraction of particles such as neutrons or electrons and hence we are able to infer the existence of wave particle duality.
ii) Bragg diffraction: This is diffraction from a multiple slits, and is similar to what occurs when waves are scattered from a periodic structure such as atoms in a crystal or rulings on a diffraction grating. Bragg diffraction is used in X-ray crystallography to deduce the structure of a crystal from the angles at which the X-rays are diffracted from it.

Dispersion is the separation of a wave into its constituent wavelengths due to interaction with a material occurring in a wavelength-dependent manner (as in thin-film interference for instance).

Wave **refraction** is a change in direction of a wave due to a change in its speed. This most commonly occurs when a wave passes from one material to another, such as a light ray passing from air into water or glass. However, light is only one example of refraction; any type of wave can undergo refraction. Another example would be physical waves passing from water into oil. At the boundary of the two media, the wave velocity is altered, the direction changes, and the wavelength increases or decreases. However, the frequency remains constant.

Snell's Law describes how light bends, or refracts, when traveling from one medium to the next. It is expressed as

$$n_1 \sin\theta_1 = n_2 \sin\theta_2$$

where n_i represents the index of refraction in medium i, and θ_i represents the angle the light makes with the normal in medium i.

Problem: The index of refraction for light traveling from air into an optical fiber is 1.44. (a) In which direction does the light bend? (b) What is the angle of refraction inside the fiber, if the angle of incidence on the end of the fiber is 22°?

Solution: (a) The light will bend toward the normal since it is traveling from a rarer region (lower n) to a denser region (higher n).

(b) Let air be medium 1 and the optical fiber be medium 2:

$$n_1 \sin\theta_1 = n_2 \sin\theta_2$$
$$(1.00)\sin 22° = (1.44)\sin\theta_2$$
$$\sin\theta_2 = \frac{1.00}{1.44}\sin 22° = (.6944)(.3746) = 0.260$$
$$\theta_2 = \sin^{-1}(0.260) = 15°$$

The angle of refraction inside the fiber is $15°$.

Light travels at different speeds in different media. The speed of light in a vacuum is represented by

$$c = 2.99792458 \times 10^8 \, m/s$$

but is usually rounded to

$$c = 3.00 \times 10^8 \, m/s.$$

Light will never travel faster than this value. The **index of refraction**, *n*, is the amount by which light slows in a given material and is defined by the formula

$$n = \frac{c}{v}$$

where *v* represents the speed of light through the given material.

Problem: The speed of light in an unknown medium is measured to be $1.24 \times 10^8 \, m/s$. What is the index of refraction of the medium?

Solution:

$$n = \frac{c}{v}$$

$$n = \frac{3.00 \times 10^8}{1.24 \times 10^8} = 2.42$$

Referring to a standard table showing indices of refraction, we would see that this index corresponds to the index of refraction for diamond.

Reflection is the change in direction of a wave at an interface between two dissimilar media such that the wave returns into the medium from which it originated. The most common example of this is light waves reflecting from a mirror, but sound and water waves can also be reflected. The law of reflection states that the angle of incidence is equal to the angle of reflection.

Reflection may occur whenever a wave travels from a medium of a given refractive index to another medium with a different index. A certain fraction of the light is reflected from the interface and the remainder is refracted. However, when the wave is moving from a dense medium into one less dense, that is the refractive index of the first is greater than the second, a critical angle exists which will create a phenomenon known as **total internal reflection**. In this situation all of the wave incident at an angle greater than the critical angle is reflected. When a wave reflects off a more dense material (higher refractive index) than that from which it originated, it undergoes a 180° phase change. In contrast, a less dense, lower refractive index material will reflect light in phase.

Fiber optics makes use of the phenomenon of total internal reflection. The light traveling through a fiber reflects off the walls at angles greater than the critical angle and thus keeps the wave confined to the narrow fiber.

Linear polarizing filters have the effect they do because of their chemical makeup. They are composed of long-chain molecules aligned in the same direction. As unpolarized light strikes the filter, the portion of the light waves, or electromagnetic vibrations, that are aligned parallel to the direction of the molecules are absorbed.

The alignment of these molecules creates a polarization axis that extends across the length of the filter. Only those vibrations that are parallel to the axis are allowed to pass through; other vibrations are blocked.

Example: A polarizing filter with a horizontal axis will allow the portion of the light waves that are aligned horizontally to pass through the filter and will block the portion of the light waves that are aligned vertically. One-half of the light is being blocked or conversely, one-half of the light is being absorbed. The image being viewed is not distorted but dimmed.

Example: If two filters are used, one with a horizontal axis and one with a vertical axis, all light will be blocked.

Skill 22.4 Use ray diagrams to analyze systems of lenses and mirrors.

Ray diagrams are a convenient way to visualize the propagation of waves and to perform reasonably accurate calculations of the effect of mirrors and lenses on light.

For mirrors, the focal point is either real (for concave mirrors) or virtual (for convex mirrors). In either case, the focal point is found by looking at the behavior of two parallel rays incident upon the mirror.

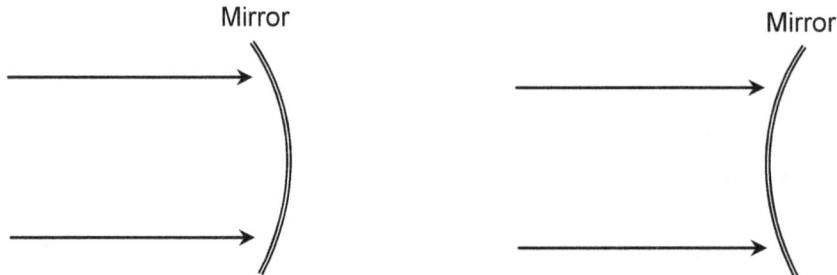

For each incident ray, the angle of reflection θ_r is equal to the angle of incidence θ_r, where the angle is measured from the normal to the mirror surface at the point of incidence.

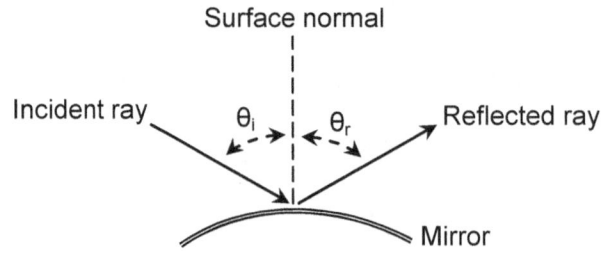

When this law is applied to the rays for either mirror, the focal points (f) are revealed as the (real or virtual) intersection of the rays. The virtual focus point is the intersection for the reflected rays when they are extended beyond the surface of the mirror.

The focal point of a lens is found in a similar manner, with the exception that instead of using the law of reflection, refraction by way of Snell's law must be applied. Since real lenses have a finite thickness, Snell's law must be used for the ray both as it enters the lens material and as it exits the lens material.

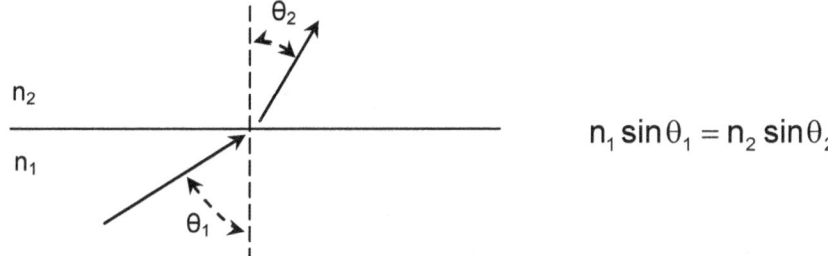

$$n_1 \sin\theta_1 = n_2 \sin\theta_2$$

As with reflection, refraction requires consideration of the normal to the surface. Also, the refractive indices must be used. It is assumed here that the refractive index of the outer medium is n_1 and the refractive index of the lens is n_2, and that $n_1 < n_2$. The intersection of parallel incident rays determines the focal point, which may again be real or virtual. Real focal points occur on the opposite side of the lens from the source of illumination, and virtual focal points occur on the same side as the source of illumination.

Only one lens case is shown here, but the principles apply equally to all variations of lenses.

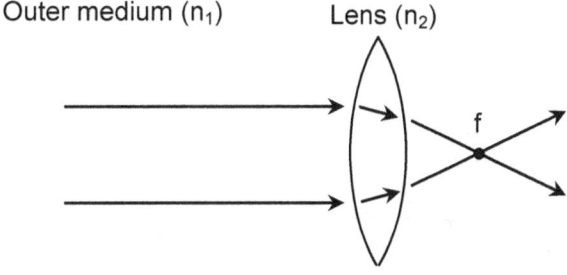

Care must be taken in properly identifying the normal and in applying Snell's law to both incidences of refraction for each ray.

Determining the point of image formation for mirrors and lenses is accomplished in a similar manner to that of the focal points. As with the focal points, image points may be either real or virtual, depending on the characteristics of the mirror or lens. To find the image point, two rays of differing angles must be traced as they interact with the lens or mirror. The initial directions of the rays can be chosen arbitrarily, but it is ideal to choose the directions such that the difficulty with determining the direction of the reflected or refracted ray is minimized. The intersection of these two rays is the image point. Only two examples are shown here, but the principles behind these examples may be applied to any variation of the situations, as well as to any combination of lenses and mirrors.

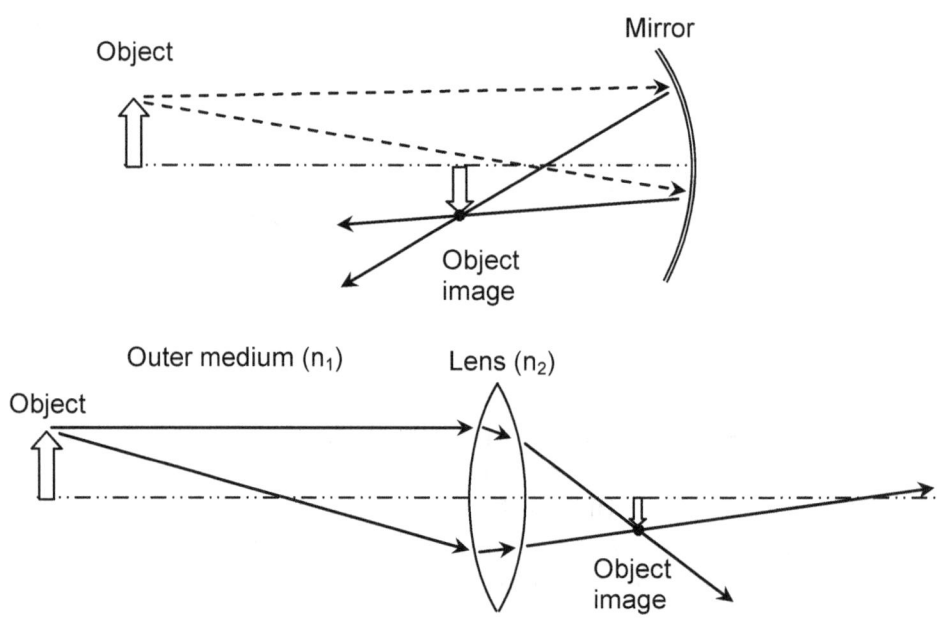

SUBAREA VI. HEAT, ELECTRICITY, MAGNETISM, AND MODERN PHYSICS

COMPETENCY 23.0 UNDERSTAND AND APPLY KNOWLEDGE OF THE PRINCIPLES OF THERMODYNAMICS.

Skill 23.1 Apply basic concepts of heat and temperature as they relate to temperature measurement and temperature-dependent properties of matter.

Heat is the thermal energy a body has due to the kinetic and potential energy of its atoms and molecules and is measured in the same units as any other form of energy, the SI unit being Joule. The traditional unit for the measurement of heat is the calorie that is related to the Joule through the relationship **1 calorie = 4.184 Joule**. Many other forms of energy, mechanical energy when you rub your hands together or electrical energy from a light bulb for instance, can be converted into heat energy. For a detailed discussion of heat energy see section IV.4.

Heat is generally measured in terms of temperature, a measure of the average internal energy of a material. Temperature is an **intensive property**, meaning that it does not depend on the amount of material. Heat content is an **extensive property** because more material at the same temperature will contain more heat. The relationship between the change in heat energy of a material and the change in its temperature is given by $\Delta Q = mC\Delta T$, where ΔQ is the change in heat energy, m is the mass of the material, ΔT is the change in temperature and C is the specific heat which is characteristic of a particular material.

There are four generally recognized temperature scales, Celsius, Fahrenheit, Kelvin and Rankine. The Kelvin and Rankine scales are absolute temperature scales corresponding to the Celsius and Fahrenheit scales, respectively. Absolute temperature scales have a zero reading when the temperature reaches absolute zero (the theoretical point at which no thermal energy exists). The absolute temperature scales are useful for many calculations in chemistry and physics.

To convert between Celsius and Fahrenheit, use the following relationship:

$$x \,°F = (5/9)(x - 32) \,°C$$

To convert to the absolute temperature scales, use the appropriate conversion below:

$$x \,°F = x + 459.67 \,°R$$

$$x \,°C = x + 273.15 \,K$$

Note that the size of each degree on the Fahrenheit/ Rankine scale is smaller than the size of a degree on the Celsius/Kelvin scale.

Skill 23.2 Apply the laws of thermodynamics to problems involving temperature, work, heat, energy, and entropy.

The first law of thermodynamics is a restatement of conservation of energy, i.e. the principle that energy cannot be created or destroyed. It also governs the behavior of a system and its surroundings. The change in heat energy supplied to a system (Q) is equal to the sum of the change in the internal energy (U) and the change in the work (W) done by the system against internal forces.

The internal energy of a material is the sum of the total kinetic energy of its molecules and the potential energy of interactions between those molecules. Total kinetic energy includes the contributions from translational motion and other components of motion such as rotation. The potential energy includes energy stored in the form of resisting intermolecular attractions between molecules. Mathematically, we can express the relationship between the heat supplied to a system, its internal energy and work done by it as

$$\Delta Q = \Delta U + \Delta W$$

Let us examine a sample problem that relies upon this law.

Problem: A closed tank has a volume of 40.0 m³ and is filled with air at 25°C and 100 kPa. We desire to maintain the temperature in the tank constant at 25°C as water is pumped into it. How much heat will have to be removed from the air in the tank to fill the tank ½ full?

Solution: The problem involves isothermal compression of a gas, so ΔU_{gas}=0. Consulting the equation above, $\Delta Q = \Delta U + \Delta W$, it is clear that the heat removed from the gas must be equal to the work done by the gas.

$$Q_{gas} = W_{gas} = P_{gas}V_1 \ln\left(\frac{V_2}{V_T}\right) = P_{gas}V_T \ln\left(\frac{\frac{1}{2}V_T}{V_T}\right) = P_{gas}V_T \ln \frac{1}{2}$$

$$= (100kPa)(40.0m^3)(-0.69314) = -2772.58 kJ$$

Thus, the gas in the tank must lose 2772.58 kJ to maintain its temperature.

To understand the second law of thermodynamics, we must first understand the concept of entropy. Entropy is the transformation of energy to a more disordered state and is the measure of how much energy or heat is available for work. The greater the entropy of a system, the less energy is available for work. The simplest statement of the second law of thermodynamics is that the entropy of an isolated system not in equilibrium tends to increase over time. The entropy approaches a maximum value at equilibrium. Below are several common examples in which we see the manifestation of the second law.

- The diffusion of molecules of perfume out of an open bottle
- Even the most carefully designed engine releases some heat and cannot convert all the chemical energy in the fuel into mechanical energy
- A block sliding on a rough surface slows down
- An ice cube sitting on a hot sidewalk melts into a little puddle; we must provide energy to a freezer to facilitate the creation of ice

When discussing the second law, scientists often refer to the "arrow of time". This is to help us conceptualize how the second law forces events to proceed in a certain direction. To understand the direction of the arrow of time, consider some of the examples above; we would never think of them as proceeding in reverse. That is, as time progresses, we would never see a puddle in the hot sun spontaneously freeze into an ice cube or the molecules of perfume dispersed in a room spontaneously re-concentrate themselves in the bottle. The above-mentioned examples are **spontaneous** as well as **irreversible**, both characteristic of increased entropy. Entropy change is zero for a **reversible process**, a process where infinitesimal quasi-static changes in the absence of dissipative forces can bring a system back to its original state without a net change to the system or its surroundings. All real processes are irreversible. The idea of a reversible process, however, is a useful abstraction that can be a good approximation in some cases.

The second law of thermodynamics may also be stated in the following ways:
1. No machine is 100% efficient.
2. Heat cannot spontaneously pass from a colder to a hotter object.

If we consider a **heat engine** that absorbs heat Q_h from a hot reservoir at temperature T_h and does work W while rejecting heat Q_c to a cold reservoir at a lower temperature T_c, $Q_h - Q_c = W$. The efficiency of the engine is the ratio of the work done to the heat absorbed and is given by

$$\varepsilon = \frac{W}{Q_h} = \frac{Q_h - Q_c}{Q_h} = 1 - \frac{Q_c}{Q_h}$$

It is impossible to build a heat engine with 100% efficiency, i.e. one where $Q_c = 0$.

Carnot described an ideal reversible engine, the **Carnot engine**, that works between two heat reservoirs in a cycle known as the **Carnot cycle** which consists of two isothermal (12 and 34) and two adiabatic processes (23 and 41) as shown in the diagram below.

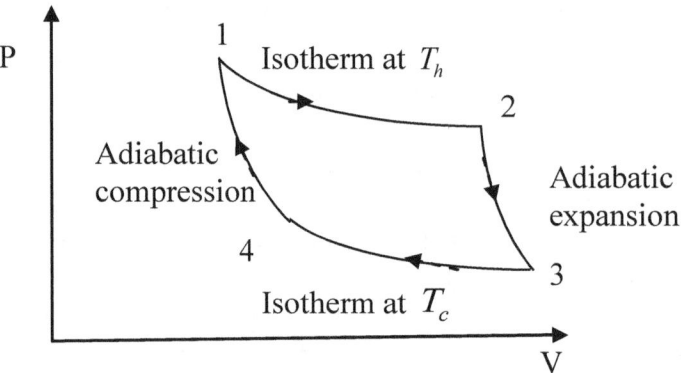

The efficiency of a Carnot engine is given by $\varepsilon = 1 - \dfrac{T_c}{T_h}$ where the temperature values are absolute temperatures. This is the highest efficiency that any engine working between T_c and T_h can reach.

According to **Carnot's theorem**, no engine working between two heat reservoirs can be more efficient than a reversible engine. All such reversible engines have the same efficiency.

The third law of thermodynamics deals with the impossibility of reaching a temperature of absolute zero and a perfectly ordered system. One common way of stating the third law of thermodynamics is, "The entropy of a perfect crystal of an element at the absolute zero of temperature is zero." The law states that, as a system approaches absolute zero, the system will become more orderly. Derivative from this law is the fact that the entropy of a given system near absolute zero is dependent only upon the temperature. At a temperature of absolute zero, there is no thermal energy or heat, so there is no molecular motion. When none of the atoms of a perfectly ordered crystal moves at all, there can be no disorder, no different possible states, and no entropy. The law also provides a reference point for all other entropy values of the system. The standard value of S^0, the entropy of a substance, is actually the integral of entropy from zero to the standard temperature 298.15 K. The values of standard entropies of elements and compounds at standard temperature are all based upon the third law of thermodynamics.

The zeroth law of thermodynamics generally deals with bodies in thermal equilibrium with each other and is the basis for the idea of temperature. Most commonly, the law is stated as, "If two thermodynamic systems are in thermal equilibrium with a third, they are also in thermal equilibrium with each other." Said another way, this very basic law simply states that if object A is same temperature as object B, and object C is the same temperature as object B, then object A and C are also the same temperature. Thermal equilibrium can thus be described as being transitive. Thermal equilibrium exists between two objects when neither object is changing temperature, or when the temperature of the objects is identical. The objects need not be in thermal contact with one another if one is certain that contact would not change their thermal properties.

Skill 23.3 Demonstrate knowledge of the kinetic-molecular theory and apply it to describe thermal properties and behaviors of solids, liquids and gases.

According to the kinetic-molecular theory, a material is modeled as a system of numerous point particles, whose collective composition and motion can be used to determine the properties of the material. Simplifying assumptions, such as a lack of any interatomic or intermolecular forces, allow the problem to remain mathematically manageable. Since it is extremely difficult, or impossible, to calculate the exact behavior of a large group of particles, the kinetic-molecular theory approaches the problem statistically. For example, the temperature of an object is actually a measurement of the average kinetic energy of the particles or molecules that compose it. As such, increasing or decreasing the kinetic energy leads to a corresponding increase or decrease in the temperature of the material.

In a solid, which is at a relatively low temperature, the individual molecules have a low kinetic energy and, thus, relatively little motion. This allows for maintenance of the orderly, periodic arrangement of the molecules in the case of a crystal, for instance. As the temperature and, concomitantly, the kinetic energy are increased, the motion of the molecules becomes faster, and the bonds or arrangements that form the crystal (or other solid material) are increasingly broken. If the temperature becomes sufficiently high, the material can no longer be considered a solid, but has changed phase into either a liquid (through melting) or a gas (through sublimation). Liquids may maintain some order, for example by remaining in a limited portion of a container, but gases will lose virtually all order and will tend to be distributed throughout any accessible volume. The types of bonds or arrangements in a particular material are determined by the characteristics of the atoms or molecules contained therein. One example, silicon, which is important for semiconductor applications, forms a diamond crystal lattice with covalent bonds. Other materials may form different bonds or crystals, by way of van der Waals forces, for instance, depending on their molecular structures and characteristics.

The particular temperatures at which phase changes between solid, liquid and gas take place are determined, in part, by the pressure on the material. Pressure and temperature, in the case of an ideal gas, are related in the kinetic-molecular theory by way of the ideal gas law, $PV = nRT$. In this equation, the product of the pressure P and the volume V is equated with the product of the number of moles n in the gas, the gas constant R and the temperature T. The relationships of the solid, liquid and gaseous phases of a material, relative to temperature and pressure, can be depicted by way of a phase diagram. A hypothetical phase diagram is shown below.

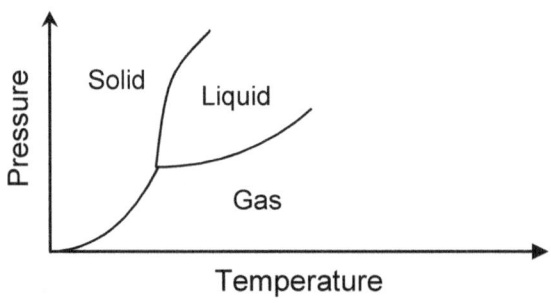

For the simple phase diagram above, the meeting point of the three curves is called the triple point. At this temperature and pressure, the phase of the material is not entirely distinguishable as either gas, liquid or solid. The particular details of the phase diagram are, again, determined by the molecular composition of the material and can generally be derived using the kinetic-molecular theory.

Skill 23.4 Analyze and solve problems involving energy, temperature, heat, and changes of state.

The **internal energy** of a material is the **sum of the total kinetic energy** of its molecules and the **potential energy** of interactions between those molecules. Total kinetic energy includes the contributions from translational motion and other components of motion such as rotation. The potential energy includes **energy stored in the form of resisting intermolecular attractions** between molecules.

The **enthalpy** (*H*) of a material is the **sum of its internal energy and the mechanical work** it can do by driving a piston. A change in the **enthalpy** of a substance is the total **energy** change caused by **adding/removing heat** at constant pressure.

When a material is heated and experiences a phase change, **thermal energy is used to break the intermolecular bonds** holding the material together. Similarly, bonds are formed with the release of thermal energy when a material changes its phase during cooling. Therefore, **the energy of a material increases during a phase change that requires heat and decreases during a phase change that releases heat**. For example, the energy of H_2O increases when ice melts and decreases when water freezes.

Heat capacity and specific heat

A substance's molar **heat capacity** is the heat required to **change the temperature of one mole of the substance by one degree**. Heat capacity has units of joules per mol- kelvin or joules per mol- °C. The two units are interchangeable because we are only concerned with differences between one temperature and another. A Kelvin degree and a Celsius degree are the same size.

The **specific heat** of a substance (also called specific heat capacity) is the heat required to **change the temperature of one gram or kilogram by one degree.** Specific heat has units of joules per gram-°C or joules per kilogram-°C.

These terms are used to solve problems involving a change in temperature by applying the formula:

$q = n \times C \times \Delta T$ where $q \Rightarrow$ heat added (positive) or evolved (negative)

$n \Rightarrow$ amount of material

$C \Rightarrow$ molar heat capacity if n is in moles, specific heat if n is a mass

$\Delta T \Rightarrow$ change in temperature $T_{final} - T_{initial}$

Example:

What is the change in energy of 10 g of gold at 25 °C when it is heated beyond its melting point to 1300 °C. You will need the following data for gold:

Solid heat capacity: 28 J/mol-K

Molten heat capacity: 20 J/mol-K

Enthalpy of fusion: 12.6 kJ/mol

Melting point: 1064 °C

Solution: First determine the number of moles used: $10 \text{ g} \times \dfrac{1 \text{ mol}}{197 \text{ g}} = 0.051 \text{ mol}$.

There are then three steps. 1) Heat the solid. 2) Melt the solid. 3) Heat the liquid. All three require energy so they will be positive numbers.

1) Heat the solid:

$$q_1 = n \times C \times \Delta T = 0.051 \text{ mol} \times 28 \frac{\text{J}}{\text{mol-K}} \times (1064 \text{ °C} - 25 \text{ °C})$$
$$= 1.48 \times 10^3 \text{ J} = 1.48 \text{ kJ}$$

2) Melt the solid: $q_2 = n \times \Delta H_{fusion} = 0.051 \text{ mol} \times 12.6 \frac{\text{kJ}}{\text{mol}}$
$$= 0.64 \text{ kJ}$$

3) Heat the liquid:

$$q_3 = n \times C \times \Delta T = 0.051 \text{ mol} \times 20 \frac{\text{J}}{\text{mol-K}} \times (1300 \text{ °C} - 1064 \text{ °C})$$
$$= 2.4 \times 10^2 \text{ J} = 0.24 \text{ kJ}$$

The sum of the three processes is the total change in energy of the gold:

$$q = q_1 + q_2 + q_3 = 1.48 \text{ kJ} + 0.64 \text{ kJ} + 0.24 \text{ kJ} = 2.36 \text{ kJ}$$
$$= 2.4 \text{ kJ}$$

A **temperature vs. heat graph** can demonstrate these processes visually. One can also calculate the specific heat or latent heat of phase change for the material by studying the details of the graph.

Example: The plot below shows heat applied to 1g of ice at -40C. The horizontal parts of the graph show the phase changes where the material absorbs heat but stays at the same temperature. The graph shows that ice melts into water at 0C and the water undergoes a further phase change into steam at 100C.

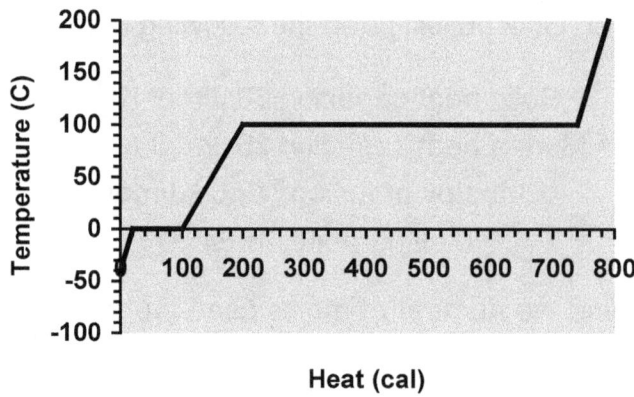

The specific heat of ice, water and steam and the latent heat of fusion and vaporization may be calculated from each of the five segments of the graph.

For instance, we see from the flat segment at temperature 0C that the ice absorbs 80 cal of heat. The latent heat L of a material is defined by the equation $\Delta Q = mL$ where ΔQ is the quantity of heat transferred and m is the mass of the material. Since the mass of the material in this example is 1g, the latent heat of fusion of ice is given by $L = \Delta Q / m = 80$ cal/g.

The next segment shows a rise in the temperature of water and may be used to calculate the specific heat C of water defined by $\Delta Q = mC\Delta T$, where ΔQ is the quantity of heat absorbed, m is the mass of the material and ΔT is the change in temperature. According to the graph, ΔQ = 200-100 =100 cal and ΔT = 100-0=100C. Thus, C = 100/100 = 1 cal/gC.

Problem: The plot below shows the change in temperature when heat is transferred to 0.5g of a material. Find the initial specific heat of the material and the latent heat of phase change.

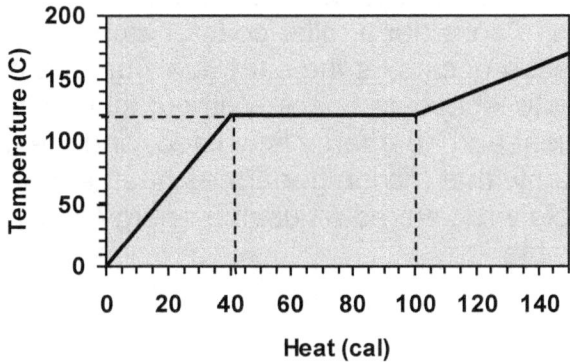

Solution: Looking at the first segment of the graph, we see that ΔQ = 40 cal and ΔT = 120 C. Since the mass m = 0.5g, the specific heat of the material is given by $C = \Delta Q / (m\Delta T)$ = 40/(0.5 X120) = 0.67 cal/gC.

The flat segment of the graph represents the phase change. Here ΔQ = 100 - 40=60 cal. Thus, the latent heat of phase change is given by $L = \Delta Q / m$ = 60/(0.5) = 120 cal/g.

Many forms of energy exist all around us. Energy is defined as the ability to do work. If you are able to measure how much work an object does, or how much heat is exchanged, you can determine the amount of energy that is in a system. Energy and work are measured in Joules. As the law of conservation of energy states, energy can be neither created nor destroyed however, it can be transformed from one form to another. Energy cannot be truly "lost" then, although energy may be wasted or not used to perform work.

Some typical forms of energy are mechanical, heat, sound, electrical, light, chemical, nuclear, and magnetic. Energy can be transformed from mechanical to heat by friction. Additionally, mechanical kinetic energy can combine with magnetic energy, to transform into electrical energy, as when a magnet is spun inside a metal coil. Electrical energy is transformed into light and heat energy when light bulb is turned on and the filament begins to glow. A firefly uses phosphorescence to transform chemical energy into light energy. Within mechanical energy, energy can transform between potential and kinetic repeatedly as is the case with a pendulum.

Energy can also be transformed into matter and vice versa. The equation $E=mc^2$ quantifies the relationship between matter and energy. The conversion of mass to other forms of energy can liberate vast amounts of energy, as shown by nuclear reactors and weapons. However, the mass equivalent of a unit of energy is very small, which is why energy loss is not typically measured by weight.

Energy transformations are classified as thermodynamically reversible or irreversible. A reversible process is one in which no energy is dissipated into empty quantum states, or states of energy with increased disorder. The easiest way to explain the concept is to consider a roller coaster car on a track. A reversible energy transformation occurs as the car travels up and down converting potential energy into kinetic and back. Without friction, the transformation is 100% efficient and no energy is wasted, and the transformation is reversible. However, we know that friction generates heat, and that the heat generated cannot be completely recovered as usable energy, which results in the transformation being irreversible.

TEACHER CERTIFICATION STUDY GUIDE

COMPETENCY 24.0   **UNDERSTAND AND APPLY KNOWLEDGE OF STATIC AND MOVING ELECTRIC CHARGES.**

Skill 24.1   **Predict the interactions between electric charges.**

Any point charge may experience force resulting from attraction to or repulsion from another charged object. The easiest way to begin analyzing this phenomenon and calculating this force is by considering two point charges. Let us say that the charge on the first point is Q_1, the charge on the second point is Q_2, and the distance between them is r. Their interaction is governed by Coulomb's Law which gives the formula for the force F as:

$$F = k\frac{Q_1 Q_2}{r^2}$$

where $k = 9.0 \times 10^9 \; \frac{N \cdot m^2}{C^2}$ (known as Coulomb's constant)

The charge is a scalar quantity, however, the force has direction. For two point charges, the direction of the force is along a line joining the two charges. Note that the force will be repulsive if the two charges are both positive or both negative and attractive if one charge is positive and the other negative. Thus, a negative force indicates an attractive force.

When more than one point charge is exerting force on a point charge, we simply apply Coulomb's Law multiple times and then combine the forces as we would in any statics problem. Let's examine the process in the following example problem.

Problem: Three point charges are located at the vertices of a right triangle as shown below. Charges, angles, and distances are provided (drawing not to scale). Find the force exerted on the point charge A.

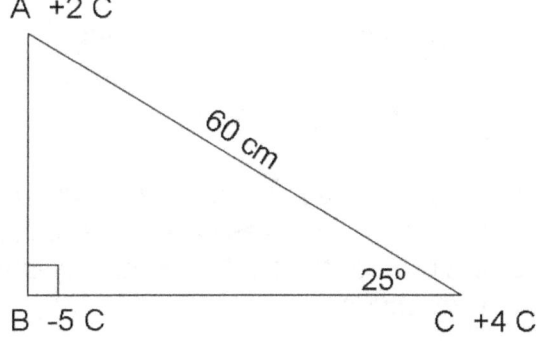

Solution: First we find the individual forces exerted on A by point B and point C. We have the information we need to find the magnitude of the force exerted on A by C.

$$F_{AC} = k\frac{Q_1 Q_2}{r^2} = 9 \times 10^9 \frac{N \cdot m^2}{C^2} \left(\frac{4C \times 2C}{(0.6m)^2} \right) = 2 \times 10^{11} N$$

To determine the magnitude of the force exerted on A by B, we must first determine the distance between them.

$$\sin 25° = \frac{r_{AB}}{60cm}$$
$$r_{AB} = 60cm \times \sin 25° = 25cm$$

Now we can determine the force.

$$F_{AB} = k\frac{Q_1 Q_2}{r^2} = 9 \times 10^9 \frac{N \cdot m^2}{C^2} \left(\frac{-5C \times 2C}{(0.25m)^2} \right) = -1.4 \times 10^{12} N$$

We can see that there is an attraction in the direction of B (negative force) and repulsion in the direction of C (positive force). To find the net force, we must consider the direction of these forces (along the line connecting any two point charges). We add them together using the law of cosines.

$$F_A^2 = F_{AB}^2 + F_{AC}^2 - 2F_{AB}F_{AC} \cos 75°$$
$$F_A^2 = (-1.4 \times 10^{12} N)^2 + (2 \times 10^{11} N)^2 - 2(-1.4 \times 10^{12} N)(2 \times 10^{11} N)^2 \cos 75°$$
$$F_A = 1.5 \times 10^{12} N$$

This gives us the magnitude of the net force, now we will find its direction using the law of sines.

$$\frac{\sin \theta}{F_{AC}} = \frac{\sin 75°}{F_A}$$
$$\sin \theta = F_{AC} \frac{\sin 75°}{F_A} = 2 \times 10^{11} N \frac{\sin 75°}{1.5 \times 10^{12} N}$$
$$\theta = 7.3°$$

Thus, the net force on A is 7.3° west of south and has magnitude 1.5×10^{12} N. Looking back at our diagram, this makes sense, because A should be attracted to B (pulled straight south) but the repulsion away from C "pushes" this force in a westward direction.

Skill 24.2 Interpret electric field diagrams and predict the influence of electric fields on electric charges.

An electric field is a vector field defined (roughly) by the magnitude and direction of the force on a unit electric charge. Electric field diagrams offer a graphical representation of the mathematics of electric fields and can be used to understand and predict the behavior of charges as they are influenced by the field.

The electric field diagram for a single, isolated point charge is illustrative of the general concepts of such diagrams. The field lines radiate from the point charge, either in the outward direction in the case of positive charge, or in the inward direction in the case of negative charge. These field lines have a direction, which indicates the direction of the force that would act on a charge placed in the field. Although only a few field lines are shown, there are, in reality, an infinite number of field lines representing the infinite number of possible positions for a test charge. These lines also extend to an infinite distance from the source. The diagrams below show the direction of the field along various paths, but do not include magnitude information.

More complicated electric field diagrams can also be drawn. Additionally, it is often helpful to draw the diagrams as small rays at various points, instead of continuous lines, with the direction of the ray corresponding to the direction of the field and with the length of the ray corresponding to the magnitude of the field. Charges are not necessarily present locally in field regions of interest, so they may or may not appear in an electric field diagram. Nevertheless, the same principles apply as in the simpler case of point charges.

The field diagram above shows the direction (arrow direction) and magnitude (arrow length) of the force that would act on a test charge in the field. It is thus possible to "track" the motion of a charged particle (or test charge) through the field. Although this approach to field diagrams allows for a moderately accurate prediction of the influence of the field on a charge, it must be noted that, if a charge is added to the scenario represented by the field, the field itself will also be altered. That is to say, the addition of charges to the situation must also include the addition (by superposition) of the field produced by that charge. While this does not necessarily have an extensive impact in the case of a single test charge, there may be a significant difference if several test charges are added. In such a case, the field diagram would have to be modified to take into account the fields of the charges, if an accurate prediction is desired.

Skill 24.3 Determine the electric potential due to a charge distribution and calculate the work involved in moving a point charge through a potential difference.

An **electric field** (**Skill 24.2**) exists in the space surrounding a charge. Electric fields have both direction and magnitude determined by the strength and direction in which they exhibit force on a test charge. The units used to measure electric fields are newtons per coulomb (N/C). **Electric potential** is simply the **potential energy** per unit of charge. Given this definition, it is clear that electric potential must be measured in joules per coulomb and this unit is known as a volt (J/C=V).

Within an electric field there are typically differences in potential energy. This **potential difference** may be referred to as **voltage**. The difference in electrical potential between two points is the amount of work needed to move a unit charge from the first point to the second point. Stated mathematically, this is:

$$V = \frac{W}{Q}$$

where V= the potential difference
W= the work done to move the charge
Q= the charge

We know from mechanics, however, that work is simply force applied over a certain distance. We can combine this with Coulomb's law to find the work done between two charges distance r apart.

$$W = F.r = k\frac{Q_1 Q_2}{r^2}.r = k\frac{Q_1 Q_2}{r}$$

Now we can simply substitute this back into the equation above for electric potential:

$$V_2 = \frac{W}{Q_2} = \frac{k\frac{Q_1 Q_2}{r}}{Q_2} = k\frac{Q_1}{r}$$

Let's examine a sample problem involving electrical potential.

Problem: What is the electric potential at point A due to the 2 shown charges? If a charge of +2.0 C were infinitely far away, how much work would be required to bring it to point A?

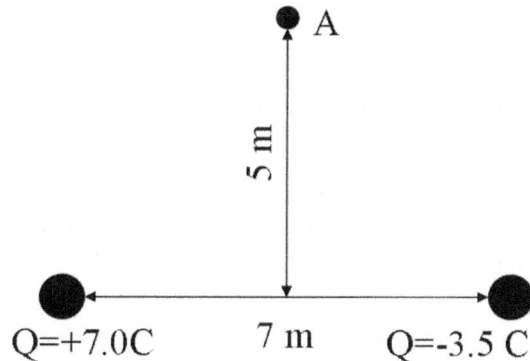

Solution: To determine the electric potential at point A, we simple find and add the potential from the two charges (this is the principle of superposition). From the diagram, we can assume that A is equidistant from each charge. Using the Pythagorean theorem, we determine this distance to be 6.1 m.

$$V = \frac{kq}{r} = k\left(\frac{7.0C}{6.1m} + \frac{-3.5C}{6.1m}\right) = 9 \times 10^9 \frac{N.m^2}{C^2}\left(0.57\frac{C}{m}\right) = 5.13 \times 10^9 V$$

Now, let's consider bringing the charged particle to point A. We assume that electric potential of these particle is initially zero because it is infinitely far away. Since now know the potential at point A, we can calculate the work necessary to bring the particle from V=0, i.e. the potential energy of the charge in the electrical field:

$$W = VQ = (5.13 \times 10^9) \times 2J = 10.26 \times 10^9 J$$

The large results for potential and work make it apparent how large the unit coulomb is. For this reason, most problems deal in microcoulombs (μC).

In any physical phenomenon, flux refers to rate of movement of a substance or energy through a certain area. Flux can be used to quantify the movement of mass, heat, momentum, light, molecules and other things. Flux depends on density of flow, area, and direction of the flow. To visualize this, imagine a kitchen sieve under a tap of flowing water. The water that passes through the sieve is the flux; the flux will decrease if we lower the water flow rate, decrease the size of the sieve, or tilt the sieve away from direction of the water's flow. Electric flux, then, is just the number of electric field lines that pass through a given area. It is given by the following equation:

$$\Phi = E(\cos\phi)A$$

where Φ = flux
E = the electric field
A = area
ϕ = the angle between the electric field and a vector normal to the surface A

Thus, if a plane is parallel to an electric field, no field lines will pass through that plane and the flux through it will be zero. If a plane is perpendicular to an electric field, the flux through it will be maximal.

Gauss's Law says that the electric flux through a surface is equal to the charge enclosed by a surface divided by a constant ε_0 (permittivity of free space). The simplest mathematical statement of this law is:

$$\Phi = Q_A/\varepsilon_0$$

where Q_A = the charge enclosed by the surface

Gauss's Law provides us with a useful and powerful method to calculate electric fields. For instance, imagine a solid conducting sphere with a net charge Q_s. We know from Gauss's Law that the electric field inside the sphere must be zero and all the excess charge lies on the outer surface of the sphere. The field produced by this sphere is the same a point charge of value Q_s. This conclusion is true whether the sphere is solid or hollow.

Skill 24.4 Determine the electric field due to a charge distribution and calculate the force on a point charge located in that electric field.

The electric field, which is defined as the force per unit charge, can be calculated for a given charge distribution by way of the principle of superposition. For a set of discrete point charges q_i, the electric field at each point **r** (the observation point) is the vector sum of the fields as contributed by each individual charge. Mathematically, this can be represented by the following expression, where **E** is the electric field, r_i is the location of the ith charge and ε_0 is the permittivity of free space (equal to 8.854 x 10^{-12} F/m).

$$\mathbf{E}(\mathbf{r}) = \frac{1}{4\pi\varepsilon_0} \sum_i \frac{q_i}{|\mathbf{r}-\mathbf{r}_i|^3} (\mathbf{r}-\mathbf{r}_i)$$

In the case of a continuous distribution of charge, an integral is required. The form of the total electric field **E**, however, is almost identical. A continuous charge distribution can be either in the form of a volume distribution (ρ) or a surface distribution (σ). For the expressions below, the location of the distribution is represented with primed coordinates (**r**′). The surface and volume over which charge is distributed are represented by S and V, respectively.

$$\mathbf{E}(\mathbf{r}) = \frac{1}{4\pi\varepsilon_0} \int_V \frac{\rho(\mathbf{r}')}{|\mathbf{r}-\mathbf{r}'|^3} (\mathbf{r}-\mathbf{r}') \, dV \quad \text{(Volume distribution)}$$

$$\mathbf{E}(\mathbf{r}) = \frac{1}{4\pi\varepsilon_0} \int_S \frac{\sigma(\mathbf{r}')}{|\mathbf{r}-\mathbf{r}'|^3} (\mathbf{r}-\mathbf{r}') \, dS \quad \text{(Surface distribution)}$$

If a particular distribution includes all three (or some combination) of the above types of charge distributions, then the total field is simply the sum of the expressions for each type of distribution.

As defined earlier, the field is a force per unit charge. As a result, the force **F** on a point charge q resulting from a particular existing charge distribution can be easily calculated according to the following formula, where **r** is the location of the point charge.

$$\mathbf{F}(\mathbf{r}) = q\, \mathbf{E}(\mathbf{r})$$

Problem: Determine the force on a point charge of -1 Coulomb, located at the origin, when point charges of 2 Coulombs each are fixed at (x,y) coordinates (-2, 0), (0, 1) and (0, -1) (in meters).

Solution: Find the field at the origin resulting from the fixed charges, then multiply by the point charge magnitude.

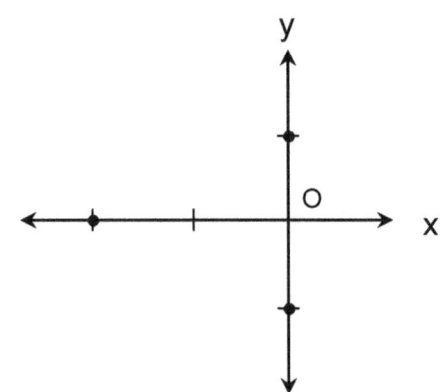

$$\mathbf{E}(O) = \frac{1}{4\pi\varepsilon_0}\left[\frac{2(0\hat{\mathbf{x}}+2\hat{\mathbf{x}})}{|0\hat{\mathbf{x}}+2\hat{\mathbf{x}}|^3} + \frac{2(0\hat{\mathbf{y}}-1\hat{\mathbf{y}})}{|0\hat{\mathbf{y}}-1\hat{\mathbf{y}}|^3} + \frac{2(0\hat{\mathbf{y}}+1\hat{\mathbf{y}})}{|0\hat{\mathbf{y}}+1\hat{\mathbf{y}}|^3}\right]\frac{C}{m^2}$$

$$\mathbf{E}(O) = \frac{2}{4\pi\varepsilon_0}\left[\frac{2\hat{\mathbf{x}}}{8} - 1\hat{\mathbf{y}} + 1\hat{\mathbf{y}}\right]\frac{C}{m^2} = \frac{\hat{\mathbf{x}}}{8\pi\varepsilon_0}\frac{C}{m^2}$$

$$\mathbf{F}_q(O) = q\mathbf{E}(O) = -1C\frac{\hat{\mathbf{x}}}{8\pi\varepsilon_0}\frac{C}{m^2} = -\hat{\mathbf{x}}\frac{1}{8\pi\varepsilon_0}\frac{C^2}{m^2}$$

Entering in the value of ε_0 gives the result in familiar units:

$$\mathbf{E}(O) = \frac{\hat{\mathbf{x}}}{8\pi \cdot 8.854\times10^{-12}\,\frac{F}{m}}\frac{C}{m^2} = \hat{\mathbf{x}}\,4.5\times10^9\,\frac{V}{m}$$

$$\mathbf{F}_q(O) = -\hat{\mathbf{x}}\frac{1}{8\pi \cdot 8.854\times10^{-12}\,\frac{F}{m}}\frac{C^2}{m^2} = -\hat{\mathbf{x}}\,4.5\times10^9\,N$$

Skill 24.5 Describe the flow of charge through different media and interpret circuit diagrams.

The flow of charge is called the current and is the rate at which electric charges pass through a conductor. A conductor is any medium that is able to carry an electric current in response to an applied difference in voltage. Inside the conductor, it is the electric field (the voltage difference/distance over which it is applied) that forces any free charges to move. In most cases, the majority of the electrons are tightly held by the atoms, but in conducting materials, electrons have a higher degree of freedom.

I. Conductors and Semiconductors

1. Metallic conductors: Metals are good conductors of current. An atom is always neutral since it has equal number of electrons and protons. The outer energy levels of the atoms of metals are not filled. There is room for the electrons to move around and from atom to atom. These free electrons require only very low energy and that is the reason metals are such good conductors of electricity. It is much easier for metals to conduct electricity.

2. Non metallic conductors: There are a number of non metallic conductors.
i) Conducting polymers: Some 20 years ago, it was discovered that polymers usually associated with insulators are very good conductors. They always need some doping with ionic components. e.g. the resistance is very low in iodine (J) doped poly-acethylene (Pac).the conduction mechanism in these compounds is along –C=C-C=C-C= chains and it is not very clear. Lot of research is going on in the field of semiconducting polymers, discovered nearly 10 years ago.

ii) Transparent conductors: Indium Tin Oxide (ITO) is the only usable transparent conductor with reasonable conductivity. It consists of Stannous Chloride doped with Indium Oxide.

iii) Ionic conductors: Solid ionic conductors are the materials behind "Ionics", including key technologies and products like primary batteries, rechargeable batteries, fuel cells, sensors etc.

iv) Specialties: Intermetallics, Silicides, Nitrides etc.Silicides are compounds of metal silicon. These are very important for microelectronics (ME) technology, but also in other applications like heating elements, Some examples are – Molybdenum Silicide, Cobalt Silicide etc.

v). Super conductors: these are a class of their own. All kinds of materials may become superconductors at low temperatures, It is not really possible to predict whether a material will become a superconductor or at what temperature.

3. Semiconductors: Semiconductors are of two types – Intrinsic, where the semiconducting properties of the material occur naturally i.e they are intrinsic to the nature of the material. Extrinsic, when the semiconducting properties of the material are manufactured by us to make the material behave in the way we want.

Almost all the materials used as semiconductors are manufactured by altering the electronic properties of the elements. The most common material used as a semiconductor is Silicon. This is achieved either by doping or junction effects (joining two materials). The purpose of doping a material is primarily to provide free electrons, which will carry current.

II. Insulators: In insulators the outer most energy levels are full. There is no room to move because the valance band is full. There is a strong resistance from the atoms to let the valence electrons move and hence no flow of charge and no current.

For more on the atomic basis of conduction see **Skill 26.6**.

Circuit diagrams: Samples of simple circuit diagrams are presented below.

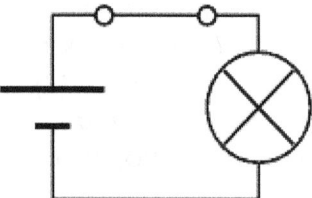 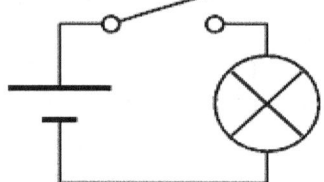 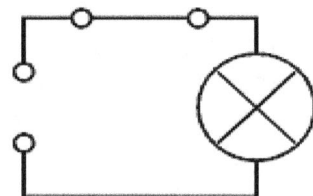

Voltage and Current
The switch is closed making a complete circuit so current can flow.

Voltage but No Current
The switch is open so the circuit is broken and current cannot flow.

No Voltage and No Current
Without the cell there is no source of voltage so current cannot flow

The diagrams above have three components – battery represented by two lines (one short and one long), a switch and lamps. Addition circuit elements and configurations are introduced in **Skill 24.6**.

Skill 24.6 Analyze AC and DC circuits composed of basic circuit elements.

Resistors and capacitors are common circuit elements and are often used together in series or parallel. Two components are in series if one end of the first element is connected to one end of the second component. The components are in parallel if both ends of one element are connected to the corresponding ends of another. A series circuit has a single path for current flow through all of its elements. A parallel circuit is one that requires more than one path for current flow in order to reach all of the circuit elements.

Below is a diagram demonstrating a simple circuit with resistors in parallel (on right) and in series (on left). Note the symbols used for a battery (noted V) and the resistors (noted R).

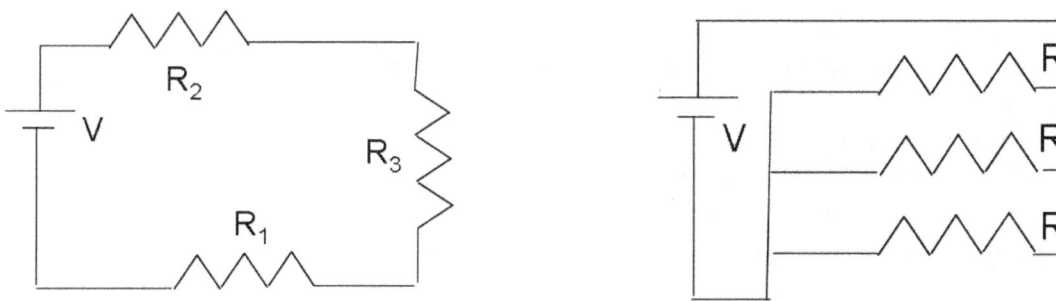

Thus, when the resistors are placed in series, the current through each one will be the same. When they are placed in parallel, the voltage through each one will be the same. To understand basic circuitry, it is important to master the rules by which the equivalent resistance (R_{eq}) or capacitance (C_{eq}) can be calculated from a number of resistors or capacitors:

Resistors in parallel: $\dfrac{1}{R_{eq}} = \dfrac{1}{R_1} + \dfrac{1}{R_2} + \cdots + \dfrac{1}{R_n}$

Resistors in series: $R_{eq} = R_1 + R_2 + \cdots + R_n$

Capacitors in parallel: $C_{eq} = C_1 + C_2 + \cdots + C_n$

Capacitors in series: $\dfrac{1}{C_{eq}} = \dfrac{1}{C_1} + \dfrac{1}{C_2} + \cdots + \dfrac{1}{C_n}$

Kirchoff's Laws are a pair of laws that apply to conservation of charge and energy in circuits and were developed by Gustav Kirchoff.

Kirchoff's Current Law: At any point in a circuit where charge density is constant, the sum of currents flowing toward the point must be equal to the sum of currents flowing away from that point.

Kirchoff's Voltage Law: The sum of the electrical potential differences around a circuit must be zero.

While these statements may seem rather simple, they can be very useful in analyzing DC circuits, those involving constant circuit voltages and currents.

Problem:
The circuit diagram at right shows three resistors connected to a battery in series. A current of 1.0 A is generated by the battery. The potential drop across R_1, R_2, and R_3 are 5V, 6V, and 10V. What is the total voltage supplied by the battery?

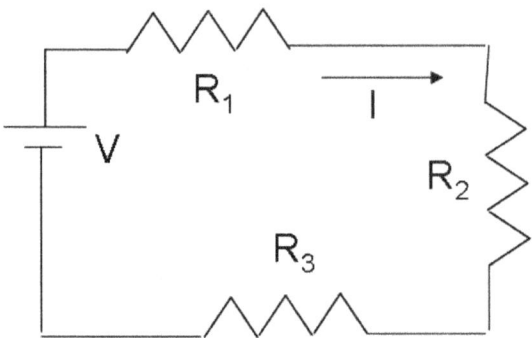

Solution:
Kirchoff's Voltage Law tells us that the total voltage supplied by the battery must be equal to the total voltage drop across the circuit. Therefore:

$$V_{battery} = V_{R_1} + V_{R_2} + V_{R_3} = 5V + 6V + 10V = 21V$$

Problem:
The circuit diagram at right shows three resistors wired in parallel with a 12V battery. The resistances of R_1, R_2, and R_3 are 4 Ω, 5 Ω, and 6 Ω, respectively. What is the total current?

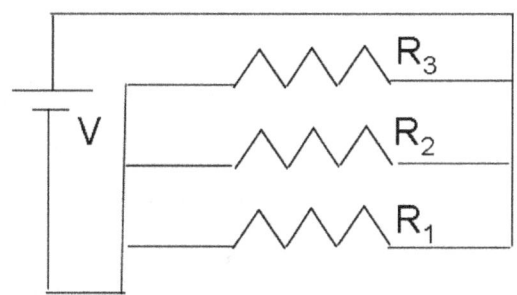

Solution:
This is a more complicated problem. Because the resistors are wired in parallel, we know that the voltage entering each resistor must be the same and equal to that supplied by the battery. We can combine this knowledge with **Ohm's Law** to determine the current across each resistor:

$$I_1 = \frac{V_1}{R_1} = \frac{12V}{4\Omega} = 3A$$

$$I_2 = \frac{V_2}{R_2} = \frac{12V}{5\Omega} = 2.4A$$

$$I_3 = \frac{V_3}{R_3} = \frac{12V}{6\Omega} = 2A$$

Finally, we use Kirchoff's Current Law to find the total current:

$$I = I_1 + I_2 + I_3 = 3A + 2.4A + 2A = 7.4A$$

In the example problems above, we have assumed that the voltage across the terminals of a battery is equal to its emf. In reality, the terminal voltage decreases slightly as the current increases. This decrease is due to the **internal resistance** r of the battery. One way to think of the internal resistance is as a small resistor in series with an ideal battery. Thus, terminal voltage drop is given by $V = \varepsilon - Ir$ where ε is the emf of the battery, r is the internal resistance and I is the current flowing through the circuit.

Resistors and capacitors may also be combined together in circuits. Below we will consider the simplest RC circuit with a resistor and a capacitor in series connected to a voltage source.

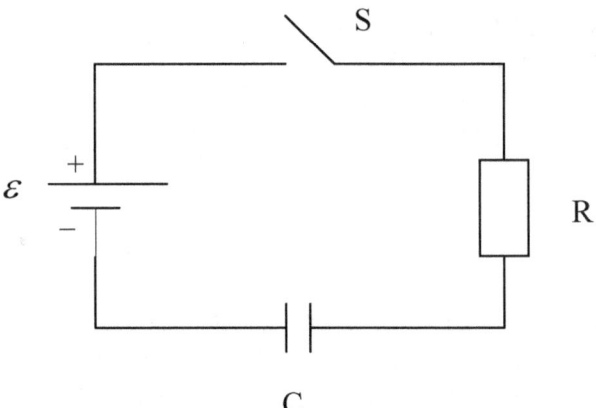

Applying Kirchhoff's first rule we get $\varepsilon = V_R + V_C = IR + \dfrac{Q}{C} = R\dfrac{dQ}{dt} + \dfrac{Q}{C}$ since the current I in the circuit is equal to the rate of increase of charge on the capacitor. If the charge on the capacitor is zero when the switch is closed at time t=0, the current I in the circuit starts out at the value ε/R and gradually falls to zero as the capacitor charge gradually rises to its maximum value. The mathematical expressions for these two quantities are

$$Q(t) = C\varepsilon(1 - e^{-t/RC}); I(t) = \dfrac{\varepsilon}{R}e^{-t/RC}$$

The product RC is known as the **time constant** of the circuit.

If the battery is now removed and the switch closed at t=0, the capacitor will slowly discharge with its charge at any point t given by

$Q(t) = Q_f e^{-t/RC}$ where Q_f is the charge that the capacitor started with at t=0.

TEACHER CERTIFICATION STUDY GUIDE

COMPETENCY 25.0   **UNDERSTAND AND APPLY KNOWLEDGE OF THE PRINCIPLES OF MAGNETISM AND INDUCED ELECTRIC FIELDS.**

Skill 25.1   **Analyze the motion of a charged particle in a magnetic field and determine the force on a current carrying conductor in a magnetic field.**

Magnetism is a phenomenon in which certain materials, known as magnetic materials, attract or repel each other. A magnet has two poles, a south pole and a north pole. Like poles repel while unlike poles attract. Magnetic poles always occur in pairs known as **magnetic dipoles**. One cannot isolate a single magnetic pole. If a magnet is broken in half, opposite poles appear at both sides of the break point so that one now has two magnets each with a south pole and a north pole. No matter how small the pieces a magnet is broken into, the smallest unit at the atomic level is still a dipole.

A large magnet can be thought of as one with many small dipoles that are aligned in such a way that apart from the pole areas, the internal south and north poles cancel each other out. Destroying this long range order within a magnet by heating or hammering can demagnetize it. The dipoles in a **non-magnetic** material are randomly aligned while they are perfectly aligned in a preferred direction in **permanent** magnets. In a **ferromagnet**, there are domains where the magnetic dipoles are aligned, however, the domains themselves are randomly oriented. A ferromagnet can be magnetized by placing it in an external magnetic field that exerts a force to line up the domains.

A magnet produces a magnetic field that exerts a force on any other magnet or current-carrying conductor placed in the field. Magnetic field lines are a good way to visualize a magnetic field. The distance between magnetic fields lines indicates the strength of the magnetic field such that the lines are closer together near the poles of the magnets where the magnetic field is the strongest. The lines spread out above and below the middle of the magnet, as the field is weakest at those points furthest from the two poles. The SI unit for magnetic field known as magnetic induction is Tesla(T) given by 1T = 1 N.s/(C.m) = 1 N/(A.m). Magnetic fields are often expressed in the smaller unit Gauss (G) (1 T = 10,000 G). Magnetic field lines always point from the north pole of a magnet to the south pole.

PHYSICS

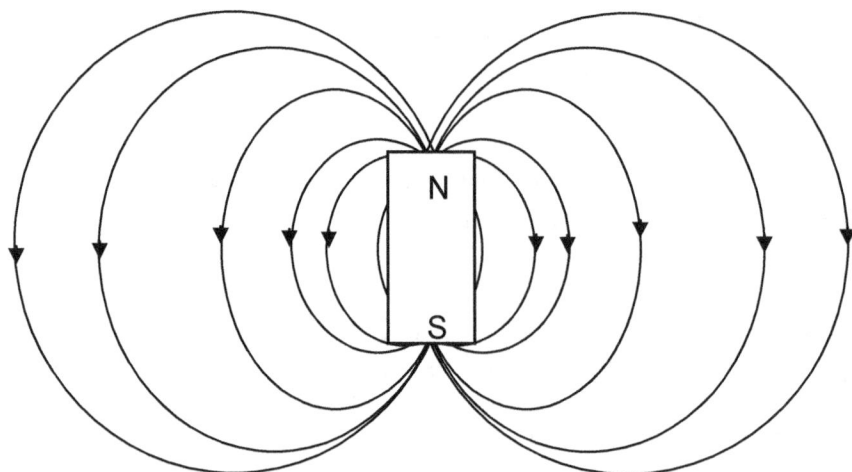

Magnetic field lines can be plotted with a magnetized needle that is free to turn in 3 dimensions. Usually a compass needle is used in demonstrations. The direction tangent to the magnetic field line is the direction the compass needle will point in a magnetic field. Iron filings spread on a flat surface or magnetic field viewing film which contains a slurry of iron filings are another way to see magnetic field lines.

The magnetic force exerted on a charge moving in a magnetic field depends on the size and velocity of the charge as well as the magnitude of the magnetic field. One important fact to remember is that only the velocity of the charge in a direction perpendicular to the magnetic field will affect the force exerted. Therefore, a charge moving parallel to the magnetic field will have no force acting upon it whereas a charge will feel the greatest force when moving perpendicular to the magnetic field.

The direction of the magnetic force, or the magnetic component of the **Lorenz force** (force on a charged particle in an electrical and magnetic field), is always at a right angle to the plane formed by the velocity vector v and the magnetic field B and is given by applying the right hand rule - if the fingers of the right hand are curled in a way that seems to rotate the v vector into the B vector, the thumb points in the direction of the force. The magnitude of the force is equal to the cross product of the velocity of the charge with the magnetic field multiplied by the magnitude of the charge.

$$F = q(v \times B) \quad or \quad F = qvB\sin(\theta)$$

Where θ is the angle formed between the vectors of velocity of the charge and direction of magnetic field.

Problem: Assuming we have a particle of 1 x 10⁻⁶ kg that has a charge of -8 coulombs that is moving perpendicular to a magnetic field in a clockwise direction on a circular path with a radius of 2 m and a speed of 2000 m/s, let's determine the magnitude and direction of the magnetic field acting upon it.

Solution: We know the mass, charge, speed, and path radius of the charged particle. Combining the equation above with the equation for centripetal force we get

$$qvB = \frac{mv^2}{r} \quad \text{or} \quad B = \frac{mv}{qr}$$

Thus B= (1 x 10⁻⁶ kg) (2000m/s) / (-8 C)(2 m) = 1.25 x 10⁻⁴ Tesla

Since the particle is moving in a clockwise direction, we use the right hand rule and point our fingers clockwise along a circular path in the plane of the paper while pointing the thumb towards the center in the direction of the centripetal force. This requires the fingers to curl in a way that indicates that the magnetic field is pointing out of the page. However, since the particle has a negative charge we must reverse the final direction of the magnetic field into the page.

A **mass spectrometer** measures the mass to charge ratio of ions using a setup similar to the one described above. m/q is determined by measuring the path radius of particles of known velocity moving in a known magnetic field.

A **cyclotron**, a type of particle accelerator, also uses a perpendicular magnetic field to keep particles on a circular path. After each half circle, the particles are accelerated by an electric field and the path radius is increased. Thus the beam of particles moves faster and faster in a growing spiral within the confines of the cyclotron until they exit at a high speed near the outer edge. Its compactness is one of the advantages a cyclotron has over linear accelerators.

The **force on a current-carrying conductor** in a magnetic field is the sum of the forces on the moving charged particles that create the current. For a current I flowing in a straight wire segment of length l in a magnetic field B, this force is given by

$$\boldsymbol{F} = I\boldsymbol{l} \times \boldsymbol{B}$$

where $\boldsymbol{l}$ is a vector of magnitude l and direction in the direction of the current.

When a current-carrying loop is placed in a magnetic field, the net force on it is zero since the forces on the different parts of the loop act in different directions and cancel each other out. There is, however, a net torque on the loop that tends to rotate it so that the area of the loop is perpendicular to the magnetic field. For a current I flowing in a loop of area A, this torque is given by

$$\tau = IA\hat{n} \times \boldsymbol{B}$$

where $\hat{n}$ is the unit vector perpendicular to the plane of the loop.

Magnetic flux (Gauss's law of magnetism)

Carl Friedrich Gauss developed laws that related electric or gravitational flux to electrical charge or mass, respectively. Gauss's law, along with others, was eventually generalized by James Clerk Maxwell to explain the relationships between electromagnetic phenomena (Maxwell's Equations).

To understand Gauss's law for magnetism, we must first define magnetic flux. Magnetic flux is the magnetic field that passes through a given area. It is given by the following equation:

$$\Phi = B(\cos\phi)A$$

where Φ = flux
B = the magnetic field
A = area
ϕ = the angle between the electric field and a vector normal to the surface A

Thus, if a plane is parallel to a magnetic field, no magnetic field lines will pass through that plane and the flux will be zero. If a plane is perpendicular to a magnetic field, the flux will be maximal.

Now we can state Gauss's law of magnetism: the net magnetic flux out of any closed surface is zero. Mathematically, this may be stated as:

$$\vec{\nabla} \cdot \vec{B} = 0$$

where $\vec{\nabla}$ = the del operator
$\vec{B}$ = magnetic field

One of the most important implications of this law is that there are no magnetic monopoles (that is, magnets always have a positive and negative pole). This is because a magnetic monopole source would give a non-zero product in the equation above. For a magnetic dipole with closed surface, of course, the net flux will always be zero. This is because the magnetic flux directed inward toward the south pole is always equal to the magnetic flux outward from the north pole.

Skill 25.2 Analyze the characteristics of magnetic fields produced by straight and coiled current-carrying conductors.

Conductors through which electrical currents travel will produce magnetic fields: The magnetic field dB induced at a distance r by an element of current Idl flowing through a wire element of length dl is given by the **Biot-Savart** law

$$dB = \frac{\mu_0}{4\pi} \frac{Idl \times \hat{r}}{r^2}$$

where μ_0 is a constant known as the permeability of free space and $\hat{r}$ is the unit vector pointing from the current element to the point where the magnetic field is calculated.

An alternate statement of this law is **Ampere's law** according to which the line integral of $B.dl$ around any closed path enclosing a steady current I is given by

$$\oint_C B \cdot dl = \mu_0 I$$

The basis of this phenomenon is the same no matter what the shape of the conductor, but we will consider three common situations:

Straight Wire
Around a current-carrying straight wire, the magnetic field lines form concentric circles around the wire. The direction of the magnetic field is given by the right-hand rule: When the thumb of the right hand points in the direction of the current, the fingers curl around the wire in the direction of the magnetic field. Note the direction of the current and magnetic field in the diagram.

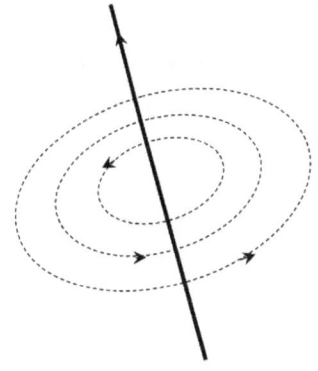

To find the magnetic field of an infinitely long (allowing us to disregarding end effects) we apply Ampere's Law to a circular path at a distance r around the wire:

$$B = \frac{\mu_0 I}{2\pi r}$$

where μ_0=the permeability of free space ($4\pi \times 10^{-7}$ T·m/A)
I=current
r=distance from the wire

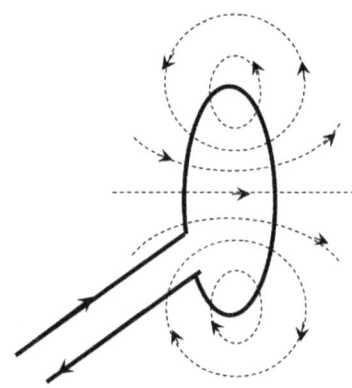

Loops
Like the straight wire from which it's been made, a looped wire has magnetic field lines that form concentric circles with direction following the right-hand rule. However, the field are additive in the center of the loop creating a field like the one shown. The magnetic field of a loop is found similarly to that for a straight wire.

In the center of the loop, the magnetic field is:

$$B = \frac{\mu_0 I}{2r}$$

Solenoids
A solenoid is essentially a coil of conduction wire wrapped around a central object. This means it is a series of loops and the magnetic field is similarly a sum of the fields that would form around several loops, as shown.

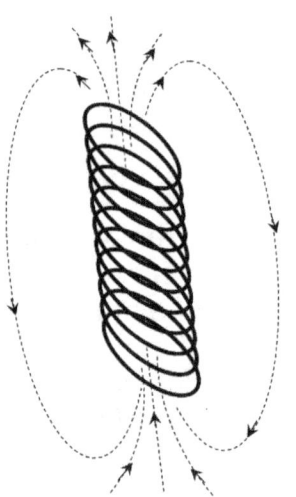

The magnetic field of a solenoid can be found as with the following equation:

$$B = \mu_0 n I$$

In this equation, n is turn density, which is simply the number of turns divided by the length of the solenoid.

Displacement current

While Ampere's law works perfectly for a steady current, for a situation where the current varies and a charge builds up (e.g. charging of a capacitor) it does not hold. Maxwell amended Ampere's law to include an additional term that includes the displacement current. This is not a true current but actually refers to changes in the electric field and is given by

$$I_d = \varepsilon_0 \frac{d\varphi_e}{dt}$$

where φ_e is the flux of the electric field. Including the displacement current, Ampere's law is given by

$$\oint \mathbf{B}.d\mathbf{l} = \mu_0 I + \mu_0 \varepsilon_0 \frac{d\varphi_e}{dt}$$

The displacement current essentially indicates that changing electric flux produces a magnetic field.

Skill 25.3 Describe and analyze the processes of electromagnetic induction.

When the magnetic flux through a coil is changed, a voltage is produced which is known as induced electromagnetic force. Magnetic flux is a term used to describe the number of magnetic fields lines that pass through an area and is described by the equation:

$$\Phi = B A \cos\theta$$

Where Φ is the angle between the magnetic field B, and the normal to the plane of the coil of area A

By changing any of these three inputs, magnetic field, area of coil, or angle between field and coil, the flux will change and an EMF can be induced. The speed at which these changes occur also affects the magnitude of the EMF, as a more rapid transition generates more EMF than a gradual one. This is described by **Faraday's law** of induction:

$$\varepsilon = -N \Delta\Phi / \Delta t$$

where ε is emf induced, N is the number of loops in a coil, t is time, and Φ is magnetic flux

The negative sign signifies **Lenz's law** which states that induced emf in a coil acts to oppose any change in magnetic flux. Thus the current flows in a way that creates a magnetic field in the direction opposing the change in flux. The right hand rule for this is that if your fingers curl in the direction of the induced current, your thumb points in the direction of the magnetic field it produces through the loop.

Consider a coil lying flat on the page with a square cross section that is 10 cm by 5 cm. The coil consists of 10 loops and has a magnetic field of 0.5 T passing through it coming out of the page. Let's find the induced EMF when the magnetic field is changed to 0.8 T in 2 seconds.

First, let's find the initial magnetic flux: Φ_i

Φ_i = BA cos θ = (.5 T) (.05 m) (.1m) cos 0° = 0.0025 T m^2

And the final magnetic flux: Φ_f

Φ_f = BA cos θ = (0.8 T) (.05 m) (.1m) cos 0° = 0.004 T m^2

The induced emf is calculated then by

ε = -N ΔΦ / Δt = - 10 (.004 T m^2 - .0025 T m^2) / 2 s = -0.0075 volts.

To determine the direction the current flows in the coil we need to apply the right hand rule and Lenz's law. The magnetic flux is being increased out of the page, with your thumb pointing up the fingers are coiling counterclockwise. However, Lenz's law tells us the current will oppose the change in flux so the current in the coil will be flowing clockwise.

Skill 25.4 Demonstrate an understanding of the operating principles of electric generators, motors and transformers

Transformers

Electromagnetic induction is used in a transformer, a device that magnetically couples two circuits together to allow the transfer of energy between the two circuits without requiring motion. Typically, a transformer consists of a couple of coils and a magnetic core. A changing voltage applied to one coil (the primary) creates a flux in the magnetic core, which induces voltage in the other coil (the secondary). All transformers operate on this simple principle though they range in size and function from those in tiny microphones to those that connect the components of the US power grid.

One of the most important functions of transformers is that they allow us to "step-up" and "step-down" between vastly different voltages. To determine how the voltage is changed by a transformer, we employ any of the following relationships:

$$\frac{V_s}{V_p} = \frac{n_s}{n_p} = \frac{I_p}{I_s}$$

where V_s=secondary voltage
V_p=primary voltage
n_s=number of turns on secondary coil
n_p=number of turns on primary coil
I_p=primary current
I_s=secondary current

Problem: If a step-up transformer has 500 turns on its primary coil and 800 turns on its secondary coil, what will be the output (secondary) voltage be if the primary coil is supplied with 120 V?

Solution:

$$\frac{V_s}{V_p} = \frac{n_s}{n_p}$$

$$V_s = \frac{n_s}{n_p} \times V_p = \frac{800}{500} \times 120V = 192V$$

Motors

Electric motors are found in many common appliances such as fans and washing machines. The operation of a motor is based on the principle that a magnetic field exerts a force on a current carrying conductor. This force is essentially due to the fact that the current carrying conductor itself generates a magnetic field; the basic principle that governs the behavior of an electromagnet. In a motor, this idea is used to convert **electrical energy into mechanical energy**, most commonly rotational energy. Thus the components of the simplest motors must include a strong magnet and a current-carrying coil placed in the magnetic field in such a way that the force on it causes it to rotate.

Motors may be run using DC or AC current and may be designed in a number of ways with varying levels of complexity. A very basic DC motor consists of the following components:

- A **field magnet**
- An **armature** with a coil around it that rotates between the poles of the field magnet
- A **power supply** that supplies current to the armature
- An **axle** that transfers the rotational energy of the armature to the working parts of the motor
- A set of **commutators** and **brushes** that reverse the direction of power flow every half rotation so that the armature continues to rotate

Generators

Generators are devices that are the opposite of motors in that they convert **mechanical energy into electrical energy**. The mechanical energy can come from a variety of sources; combustion engines, blowing wind, falling water, or even a hand crank or bicycle wheel. Most generators rely upon electromagnetic induction to create an electrical current. These generators basically consist of magnets and a coil. The magnets create a magnetic field and the coil is located within this field. Mechanical energy, from whatever source, is used to spin the coil within this field. As stated by Faraday's Law, this produces a voltage.

Skill 25.5 Identify applications of magnets and magnetic fields in technology and daily living.

Electromagnetism is the foundation for a vast number of modern technologies ranging from computers to communications equipment. More mundane technologies such as motors and generators are also based upon the principles of electromagnetism. The particular understanding of electrodynamics can either be in terms of quantum mechanics (quantum electrodynamics) or classical electrodynamics, depending on the type of phenomenon being analyzed. In classical electrodynamics, which is a sufficient approximation for most situations, the electric field, resulting from electric charge, and the magnetic field, resulting from moving charges, are the parameters of interest and are related through Maxwell's equations.

Meters

A number of different types of meters use electromagnetism or are designed to measure certain electromagnetic parameters. For example, older forms of ammeters (galvanometers), when supplied with a current, provided a measurement through the deflection of a spring-loaded needle. A coil connected to the needle acted as an electromagnet which, in the presence of a permanent magnetic field, would be deflected in the same manner as a rotor, as mentioned previously. The strength of the electromagnet, and thus the extent of the deflection, is proportional to the current. Further, the spring limits the deflection in such a manner that a reasonably accurate measurement of the current is provided.

Magnetic Media

Although the cassette tape has fallen out of favor with popular culture, magnetic media are still widely used for information storage. Magnetic strips on the back of credit and identification cards, computer hard drive disks and magnetic tapes (such as those contained in cassettes) are all examples of magnetic media. The principle underlying magnetic media is the sequential magnetization of a region of the medium. A special head is able to detect spatial magnetic fluctuations in the medium which are then converted into an electrical signal. The head is often able to "write" to the medium as well. The electrical signal from the medium can be converted into sound or video, as with the video or audio cassette player, or it can be digitized for use in a computer.

Note that motors, generators, and transformers (**Skill 25.4**) provide some of the most common examples of how technology has exploited electromagnetic phenomena.

COMPETENCY 26.0   **UNDERSTAND AND APPLY KNOWLEDGE OF THE BASIC CONCEPTS AND APPLICATIONS OF MODERN PHYSICS.**

Skill 26.1   Demonstrate knowledge of a quantum model of atomic structure (e.g., the Bohr model), including the relationship between changes in electron energy levels and atomic spectra.

In the West, the Greek philosophers Democritus and Leucippus first suggested the concept of the atom. They believed that all atoms were made of the same material but that varied sizes and shapes of atoms resulted in the varied properties of different materials. By the 19th century, John Dalton had advanced a theory stating that each element possesses atoms of a unique type. These atoms were also thought to be the smallest pieces of matter which could not be split or destroyed.

Atomic structure began to be better understood when, in 1897, JJ Thompson discovered the electron while working with cathode ray tubes. Thompson realized the negatively charged electrons were subatomic particles and formulated the "**plum pudding model**" of the atom to explain how the atom could still have a neutral charge overall. In this model, the negatively charged electrons were randomly present and free to move within a soup or cloud of positive charge. Thompson likened this to the dried fruit that is distributed within the English dessert plum pudding though the electrons were free to move in his model.

Ernest Rutherford disproved this model with the discovery of the nucleus in 1909. In **Rutherford's alpha scattering** experiments, he found that alpha particles striking a thin gold foil were scattered at large angles which indicated that the positive charge in an atom was concentrated in a small volume. Rutherford proposed a new "**planetary**" **model** of the atom in which electrons orbited around a positively charged nucleus like planets around the sun. Over the next 20 years, protons and neutrons (subnuclear particles) were discovered while additional experiments showed the inadequacy of the planetary model.

As quantum theory was developed and popularized (primarily by Max Planck and Albert Einstein), chemists and physicists began to consider how it might apply to atomic structure. Niels Bohr put forward a model of the atom in which electrons could only orbit the nucleus in circular orbitals with specific distances from the nucleus, energy levels, and angular momentums. In this model, electrons could only make instantaneous "quantum leaps" between the fixed energy levels of the various orbitals. The Bohr model of the atom was altered slightly by Arnold Sommerfeld in 1916 to reflect the fact that the orbitals were elliptical instead of round.

Though the Bohr model is still thought to be largely correct, it was discovered that electrons do not truly occupy neat, cleanly defined orbitals. Rather, they exist as more of an "electron cloud." The work of Louis de Broglie, Erwin Schrödinger, and Werner Heisenberg showed that an electron can actually be located at any distance from the nucleus. However, we can find the *probability* that the electrons exists at given energy levels (i.e., in particular orbitals) and those probabilities will show that the electrons are most frequently organized within the orbitals originally described in the Bohr model.

The quantum structure of the atom describes electrons in discrete energy levels surrounding the nucleus. When an electron moves from a high energy orbital to a lower energy orbital, a quantum of electromagnetic radiation is emitted, and for an electron to move from a low energy to a higher energy level, a quantum of radiation must be absorbed. The particle that carries this electromagnetic force is called a **photon**. The quantum structure of the atom predicts that only photons corresponding to certain wavelengths of light will be emitted or absorbed by atoms. These distinct wavelengths are measured by **atomic spectroscopy**.

In **atomic absorption spectroscopy**, a continuous spectrum (light consisting of all wavelengths) is passed through the element. The frequencies of absorbed photons are then determined as the electrons increase in energy. An **absorption spectrum** in the visible region usually appears as a rainbow of color stretching from red to violet interrupted by a few black lines corresponding to distinct wavelengths of absorption.

In **atomic emission spectroscopy**, the electrons of an element are excited by heating or by an electric discharge. The frequencies of emitted photons are then determined as the electrons release energy. An **emission spectrum** in the visible region typically consists of lines of light at certain colors corresponding to distinct wavelengths of emission. The bands of emitted or absorbed light at these wavelengths are called **spectral lines**. **Each element has a unique line spectrum**. Light from a star (including the sun) may be analyzed to determine what elements are present.

Skill 26.2 Describe types, properties, and applications of radioactivity and the effects of radioactivity on living organisms.

Radioactive decay is characterized by the emission of certain particles of various types and energies from an atom. The three most important types of radioactivity are alpha decay, beta decay, and gamma decay. Alpha decay involves the emission of a helium (^{4}He) nucleus, which is constituted by two neutrons and two protons. Beta decay involves the emission of either an electron and an antineutrino (β^- decay) or a positron and a neutrino (β^+ decay). A nucleus that undergoes a β^- decay decreases its atomic number (number of protons) by unity and increases its neutron number (number of neutrons) by unity. In the case of a β^+ decay, the nucleus increases its atomic number by unity and, correspondingly, decreases its neutron number by unity. A third type of radioactivity is gamma decay, which involves the lowering of the energy level of an atom through emission of a gamma ray (a high-energy photon).

Other types of decay may occur, but these are generally variations of the above three examples. A number of different types of nucleon emissions (of which alpha emission is one example) can occur, as can various types of beta emissions. Gamma decay can take place in a range of energies, corresponding to different frequencies of emitted photons. The rates at which these decays take place are typically characterized by a half life, or the time required for half of a given sample to undergo the radioactive decay.

Radioactive materials have been used in a number of applications, both in the technological and the biological spheres. For instance, small amounts of radioactive material can be injected or ingested in patients for diagnostic purposes. Under the appropriate form of detector, the material can highlight internal body structures and, potentially, abnormalities, thus reducing the need for invasive surgery. Also, examination of naturally occurring radioactive materials can provide, in some contexts, a method for dating past events. Certain radioactive materials have also been used for the purpose of 'tagging' fissile nuclear materials, such as are used in nuclear reactors or nuclear weapons, so as to make these dangerous materials less prone to theft.

Since alpha, beta (positron and electron) and gamma emissions are all forms of ionizing radiation, exposure can cause severe damage to living organisms, depending on the intensity of the radiation and the duration of exposure. Since these radioactive emissions are highly energetic, they can, during incidence upon biological tissue, transfer their kinetic energy to cells or DNA, thus causing damage. Damage to DNA is a particularly adverse result, as this can lead to abnormal cell functioning and, in some cases, various forms of cancer. Depending on the type of radioactivity, however, the particle flux on a living organism can be reduced through the use of shielding materials.

Skill 26.3 **Balance particle equations and solve radioactive decay problems involving half-life, energy, mass and charge.**

Some nuclei are unstable and emit particles and electromagnetic radiation. These emissions from the nucleus are known as **radioactivity.** It is often found that some isotopes of an element are radioactive, typically the ones with an excess of neutrons in the nucleus. The unstable isotopes are known as **radioisotopes**; and the nuclear reactions that spontaneously alter them are known as **radioactive decay.** Particles commonly involved in nuclear reactions are listed in the following table:

Particle	Neutron	Proton	Electron	Alpha particle	Beta particle	Gamma rays
Symbol	$_0^1 n$	$_1^1 p$ or $_1^1 H$	$_{-1}^0 e$	$_2^4 \alpha$ or $_2^4 He$	$_{-1}^0 \beta$ or $_{-1}^0 e$	$_0^0 \gamma$

Artificial or **induced radioactivity** is the production of radioactive isotopes by bombarding an element with high velocity particles such as neutrons.

In **alpha decay**, an atom emits an alpha particle. An alpha particle contains two protons and two neutrons. This makes it identical to a helium nucleus and so an alpha particle may be written as He^{2+} or it can be denoted using the Greek letter α. Because a nucleus decaying through alpha radiation loses protons and neutrons, the mass of the atom loses about 4 Daltons and it actually becomes a different element (transmutation). For instance:

$$^{238}U \rightarrow {}^{234}Th + \alpha$$

Radioactive heavy nuclei including uranium and radium typically decay by emitting alpha particles. The alpha decay often leaves the nucleus in an excited state with the extra energy subsequently removed by gamma radiation. The energy of alpha particles can be readily absorbed by skin or air and so alpha decaying substances are only harmful to living things if they are introduced internally.

Like alpha decay, **beta decay** involves emission of a particle. In this case, though, it is a beta particle, which is either an electron or positron. Note that a positron is the antimatter equivalent of an electron and so these particles are often denoted β$^-$ and β$^+$. Beta plus and minus decay occur via roughly opposite paths. In beta minus decay, a neutron is converted in a proton (specifically, a down quark is converted to an up quark), an electron and an anti-neutrino; the latter two are emitted. In beta plus decay, on the other hand, a proton is converted to a neutron, a positron, and a neutrino; again, the latter two are emitted. As in alpha decay, a nucleus undergoing beta decay is transmuted into a different element because the number of protons is altered.

However, because the total number of nucleons remains unchanged, the atomic mass remains the same (note, that the neutron is actually slightly heavier than a proton so mass is gained during beta plus decay). So examples of beta decay would be:

$$^{137}_{55}Cs \rightarrow {}^{137}_{56}Ba + e^- + \bar{v}_e \quad \text{(beta minus decay)}$$

$$^{22}_{11}Na \rightarrow {}^{22}_{10}Ne + e^+ + v_e \quad \text{(beta plus decay)}$$

Beta decaying isotopes, such as Strontium 90, are commonly used in cancer treatment. These particles are better able to penetrate skin than alpha particles and so exposure to larger amounts of beta particles poses a risk to all living things.

Gamma radiation is quite different from alpha and beta decay in that it does not involve the emission of nucleon-containing particles or the transmutation of elements. Rather, gamma-ray photons are emitted during gamma decay. These gamma rays are a specific form of electromagnetic radiation that results from certain sub-atomic particle contacts. For instance, electron-positron annihilation leads to the emission of gamma rays. More commonly, though, gamma rays are emitted by nuclei left in an excited state following alpha or beta decay. Thus, gamma decay lowers the energy level of a nucleus, but does not change its atomic mass or charge. The high energy content of gamma rays, coupled with their ability to penetrate dense materials, make them a serious risk to living things.

While the radioactive decay of an individual atom is impossible to predict, a mass of radioactive material will decay at a specific rate. Radioactive isotopes exhibit exponential decay and we can express this decay in a useful equation as follows:

$$A = A_0 e^{kt}$$

Where A is the amount of radioactive material remaining after time t, A_0 is the original amount of radioactive material, t is the elapsed time, and k is the unique activity of the radioactive material. Note that k is unique to each radioactive isotope and it specifies how quickly the material decays. Sometimes it is convenient to express the rate of decay as half-life. **A half-life is the time needed for half a given mass of radioactive material to decay.** Thus, after one half-life, 50% of an original mass will have decayed, after two half lives, 75% will have decayed and so on.

Let's examine a sample problem related to radioactive decay.

Problem: Radiocarbon dating has been used extensively to determine the age of fossilized organic remains. It is based on the fact that while most of the carbon atoms in living things is ^{12}C, a small percentage is ^{14}C. Since ^{14}C is a radioactive isotope, it is lost from a fossilized specimen at a specific rate following the death of an organism. The original and current mass of ^{14}C can be inferred from the relative amount of ^{12}C. So, if the half-life of ^{14}C is 5730 years and a specimen that originally contained 1.28 mg of ^{14}C now contains 0.10 mg, how old is the specimen?

In certain problems, we may be simply provided with the activity, k, but in this problem we must use the information given about half-life to solve for k.

Since we know that after one half-life, 50% of the material remains radioactive, we can plug into the governing equation above:

$$A = A_0 e^{kt}$$
$$0.5 A_0 = A_0 e^{5730k}$$
$$k = (\ln(0.5))/5730 = -0.0001209$$

Having determined k, we can use this same equation again to determine how old the specimen described above must be:

$$A = A_0 e^{kt}$$
$$0.10 = 1.28 e^{-0.0001209 t}$$
$$t = \frac{\ln\left(\frac{0.10}{1.28}\right)}{-0.0001209} = 21087$$

Thus, the specimen is 21,087 years old.

Note that this same equation can be used to calculate the half-life of an isotope if information regarding the decay after a given number of years were provided.

Skill 26.4 Describe the quantum mechanical nature of the interaction between radiation and matter.

Classical theories of the interaction of electromagnetic radiation (light) and matter are often sufficient for an adequate description of many macroscopic phenomena. In classical electrodynamics, for instance, matter can often be treated as a continuous medium described by parameters such as permittivity, permeability and conductivity, and light can be treated as a wave. For microscopic phenomena, however, quantum mechanics provides the foundation for a much more satisfactory characterization, with light being modeled as particles (that may exhibit some wave characteristics) of discrete energies, rather than being modeled purely as a wave with a continuous range of available energies.

Particles of light (photons) are the mediating particle (boson) for the electromagnetic interaction. According to quantum electrodynamics, particles of matter that have a charge (such as electrons) are particles that are capable of coupling with photons. As a result, two charged particles may exchange photons (and, thus, energy), causing a deflection in the paths of the particles. Macroscopically, when this takes place for a large number of particles in a material, this is manifested as an electric or magnetic force. For the microscopic scale, the interaction of two charged particles can be depicted using a Feynman diagram. The Feynman diagram below depicts the deflection ("repulsion") of two electrons due to an exchange of a photon.

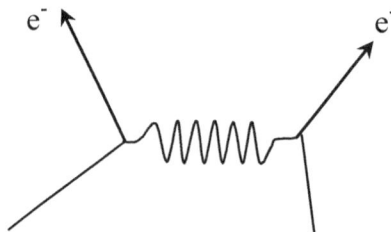

Since photons only couple with charged particles, neutral particles do not experience any electromagnetic forces. This makes neutral particles, such as neutrinos, difficult to detect.

A quintessential phenomenon demonstrating the usefulness of quantum theory in describing the interaction between light and matter is that of the photoelectric effect. When light above a certain frequency (energy) is incident upon a metallic material, electrons are ejected from the surface with a specific kinetic energy. This threshold energy cannot be explained classically, and the use of the concept of the photon in the explanation of the photoelectric effect was a key step towards the development of quantum theory.

Additionally, unlike classical mechanics, which describes the interaction of light and matter as purely deterministic in nature, quantum mechanics describes the interaction probabilistically in terms of a wave function. Both particles of matter (such as electrons) and particles of light (photons) have associated wave functions, the squares of which are probability distributions. As a result, the electromagnetic interaction is, from a quantum mechanical perspective, a phenomenon that must be described in terms of probabilities of events. That is, in the case of the interaction between two particles by way of a photon, for instance, the behavior of the particles and the photon can only be predicted in terms of probabilities.

Skill 26.5 Describe the wave-particle duality of radiation and matter.

The dual wave and particle nature of light has long been considered. In 1924 **Louis de Broglie** suggested that not only light but all matter, particularly electrons, may exhibit wave as well as particle behavior. He proposed that the frequency f and wavelength λ of electron waves are given by the equations

$$f = \frac{E}{h}; \lambda = \frac{h}{p}$$

where p is the momentum of the electron, E is its energy and h is Planck's constant. These are the same relations that Planck proposed for photons. Using deBroglie's equations and considering electrons as standing waves in a circular Bohr orbit, the discrete energy states of an electron could be explained and led to the same set of energy levels found by Bohr. Schrodinger developed these ideas into **wave mechanics**, a general method for finding the quantization condition for a system.

Wave-particle duality is also expressed by **Heisenberg's uncertainty principle** which places a limit on the accuracy with which one can measure the properties of a physical system. This limit is not due to the imperfections of measuring instruments or experimental methods but arises from the fundamental wave-particle duality inherent in quantum systems.

One statement of the uncertainty principle is made in terms of the position and momentum of a particle. If Δx is the uncertainty in the position of a particle in one dimension and Δp the uncertainty in its momentum in that dimension, then according to the uncertainty principle

$$\Delta x \Delta p \geq \hbar / 2$$

where the reduced Planck's constant $\hbar = h/2\pi = 1.05457168 \times 10^{-34} J.s$

Thus if we measure the position of a particle with greater and greater accuracy, at some point the accuracy in the measurement of its momentum will begin to fall. A simple way to understand this is by considering the wave nature of a subatomic particle. If the wave has a single wavelength, then the momentum of the particle is also exactly known using the DeBroglie momentum-wavelength relationship. The position of the wave, however, extends through all space. If waves of several different wavelengths are superposed, the position of the wave becomes increasingly localized as more wavelengths are added. The increased spread in wavelength, however, then results in an increased momentum spread.

An alternate statement of the uncertainty principle may be made in terms of energy and time.

$$\Delta E \Delta t \geq \hbar/2$$

Thus, for a particle that has a very short lifetime, the uncertainty in the determination of its energy will be large.

Problem: If a proton is confined to a nucleus that is approximately $10^{-15}m$ in diameter, estimate the minimum uncertainty in its momentum in any one dimension.

Solution: The uncertainty of the position of the proton in any dimension cannot be greater than $10^{-15}m$. Using the uncertainty principle we find that the approximate uncertainty in its momentum in any one dimension must be greater than
$\Delta p = \hbar/(2\Delta x) \approx 10^{-19} Kg.m/s$

Skill 26.6 Describe the quantum mechanical electron properties of conductors, semiconductors, and insulators.

Conductors, semiconductors and insulators are each characterized by their respective conductivities, which describe the ease with which electrons flow through the material. An accurate prediction of the conductivity for a given material can be obtained by way of quantum theory.

By solving the Schrödinger equation for a (nearly) free electron in a periodic potential, it can be found that the available energy states for the electron exist not as a continuous range of energies, but as a set of discrete energy bands. These energy bands each contain a finite number of available states that may be occupied at a given time by an electron. Given that systems tend to minimize their energies, the electrons in a material tend to reside in the lowest available energy states, unless they are excited by some impartation of energy. Since a real material actually contains a large number of electrons, this means that the lower-energy bands are generally filled, while the higher-energy bands are generally unfilled. The band diagram below illustrates a hypothetical example.

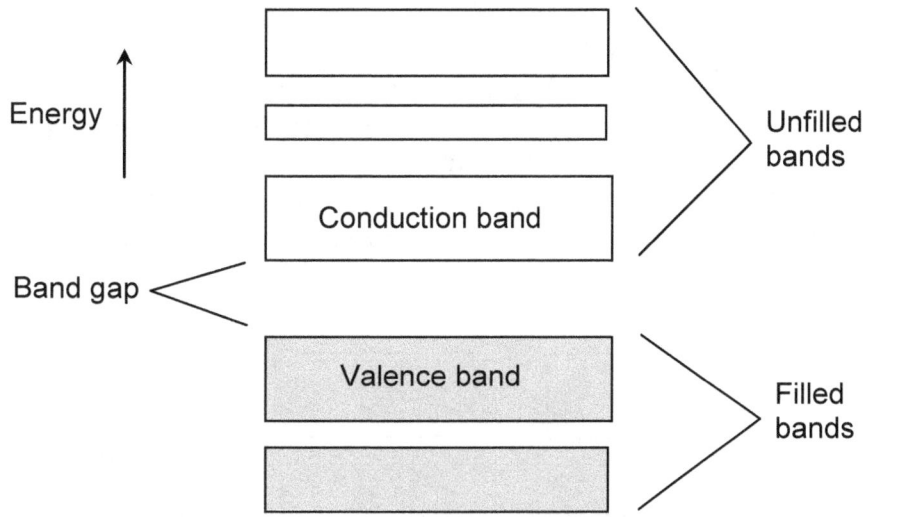

In light of the band structure of energy levels for electrons in a material, the conductivities of conductors, semiconductors and insulators may be explained quantum mechanically. Since electrons in the valence band are not free to move (in the sense of a net current), a material in an absolute ground state should display little or no conductivity. Conduction (movement) of charge requires excitation of electrons into one of the conduction bands. This may be the result of an applied external field or the result of thermal heating. In the former case, incident photons can be absorbed by electrons, causing them to move into a higher energy state in a conduction band. In the latter case, electrons may absorb phonons (a vibrational mode that acts like a particle), also causing them to move into a higher energy state. These electrons are then "free" to move about in the material, and, in the presence of an applied field, this leads to an electric current. The degree to which there are free charge carriers that can act as a current in an applied field and at a given temperature is, essentially, the conductivity of the material.

In the case of semiconductors and insulators, the band structure is such that there is a gap between the highest-energy valence electrons and the lowest-energy conduction band. The above diagram depicts this scenario. The difference between semiconductors and insulators is simply a matter of degree: semiconductors have a smaller band gap, whereas insulators have a larger band gap. As the temperature increases, there is a commensurate increase in the number of free electrons in the conduction band. While this effect increases the conductivity to an extent, the increase is limited by the thermal motion of the nuclei in the crystal lattice, which impedes the flow of electrons. Thus, there are two competing phenomena occurring in an insulator or semiconductor as temperature increases, therefore limiting the ability to increase the conductivity in this manner.

For conductors, the energy band that contains the highest-energy valence electrons is only partially filled. As a result, there are a number of free conduction states within that very energy band, and electrons need little additional energy to move into these free states. As a result, in a conductor there are a vast number of free electrons, and the conductivity is commensurately high. The band diagrams below compare the characteristics of a hypothetical conductor, semiconductor and insulator.

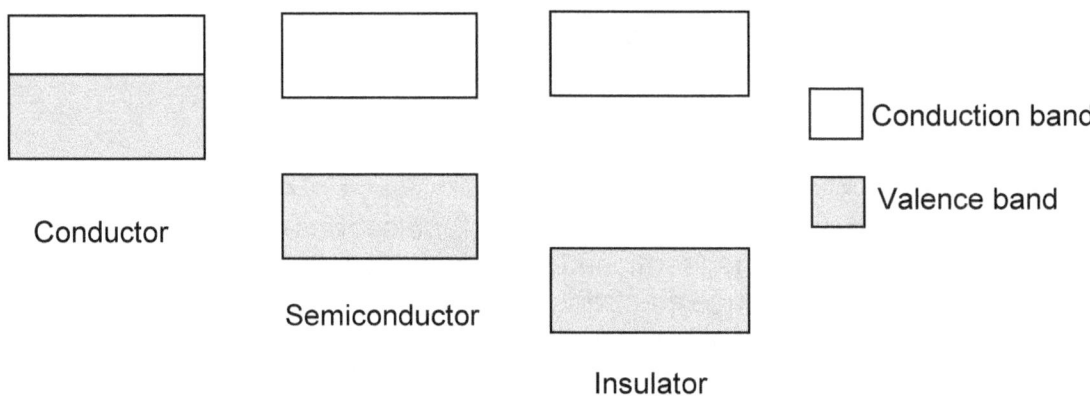

Skill 26.7 Apply the concepts of special relativity as they relate to time, space and mass.

Postulates of special relativity

Einstein's theory of special relativity was built upon the foundation of Galilean relativity and incorporated an analysis of electromagnetics. Galilean relativity is based upon the concept that the laws of physics are invariant with respect to inertial frames of reference; this statement of invariance is one of the two fundamental postulates of special relativity. In the context of classical electrodynamics, it is found that Maxwell's equations do not imply any variation in the speed of light with respect to the relative motion of the source and observer (i.e., the inertial frame of reference). As a result, the second fundamental postulate of special relativity is a statement of the invariance of the speed of light, c, with respect to inertial reference frames. The postulates of special relativity are given concisely below.

1. Special principle of relativity: The laws of physics are same in all inertial frames of reference.
2. Invariance of c: The speed of light in a vacuum is a universal constant for all observers, regardless of the inertial frame of reference or the relative motion of the source.

Space-time

The implications of these two postulates can be seen through a comparison of the various parameters as viewed from two different inertial frames of reference, S and S'. Let frame S be stationary and frame S' have a velocity v along the x-axis. The Lorentz transformations require that an event at time t and position (x, y, z) in frame S must occur at time t' and position (x', y', z') in frame S'.

$$t' = \gamma \left(t - \frac{vx}{c^2} \right)$$
$$x' = \gamma (x - vt)$$
$$y' = y$$
$$z' = z$$

Here, the Lorentz factor γ is defined as follows.

$$\gamma = \frac{1}{\sqrt{1 - \frac{v^2}{c^2}}}$$

The Lorentz transformation equations shows that, contrary to classical mechanics, where t' = t, time and space are codependent in the context of special relativity. As a result, rather than referring to time and space as separate aspects of the universe, physicists will often speak of space-time, a singular characteristic of the universe that is manifested in both spatial and temporal dimensions. The following expression, which is invariant with respect to the Lorentz transformation, has the characteristics of the "length" of a vector in space-time.

$$x^2 + y^2 + z^2 - c^2 t^2$$

The interconnectedness of space and time (i.e., space-time) results in counterintuitive phenomena such as length contraction and time dilation. Furthermore, the concept of simultaneity becomes relative, as time is a function of the relative velocities of inertial frames of reference. In daily life, since the objects encountered are traveling at speeds much less than c, relativistic effects are not seen. For these circumstances, the Lorentz transformations reduce approximately to the classical transformations, where all dimensions in frame S are the same as those in frame S', with the exception of x' = x − vt.

Mass and energy

The constancy of the speed of light also has implications with regard to mass and energy. The relativistic energy of an object is given below.

$$E^2 = (pc)^2 + (mc^2)^2$$

Here, p is the momentum of the object. This equation shows that energy and mass are not distinct parameters, but are related. Furthermore, since energy and momentum are dependent upon the inertial frame of reference, mass is likewise dependent upon the inertial frame of reference. The mass of an object, m, varies with velocity according to the Lorentz factor, as given below. The rest mass m_0 is used here.

$$m = \gamma m_0$$

As a result of the above expression, as the speed of an object approaches the speed of light and as the Lorentz factor commensurately approaches infinity, the mass of the object also approaches infinity. Consequently, the amount of energy required to further accelerate the object approaches infinity. This all leads to the concept of a "universal speed limit": no object may travel faster than the speed of light. Thus, it can be seen that the constancy of the speed of light leads to unique, and sometimes difficult to comprehend, relationships between space, time and mass, such as are not seen in the simpler case of classical mechanics.

TEACHER CERTIFICATION STUDY GUIDE

Sample Test

DIRECTIONS: Read each item and select the best response.

1. Which statement best describes a valid approach to testing a scientific hypothesis?
   (Easy)

   A. Use computer simulations to verify the hypothesis

   B. Perform a mathematical analysis of the hypothesis

   C. Design experiments to test the hypothesis

   D. All of the above

2. Which description best describes the role of a scientific model of a physical phenomenon?
   (Average Rigor)

   A. An explanation that provides a reasonably accurate approximation of the phenomenon

   B. A theoretical explanation that describes exactly what is taking place

   C. A purely mathematical formulation of the phenomenon

   D. A predictive tool that has no interest in what is actually occurring

3. Which situation calls might best be described as involving an ethical dilemma for a scientist?
   (Rigorous)

   A. Submission to a peer-review journal of a paper that refutes an established theory

   B. Synthesis of a new radioactive isotope of an element

   C. Use of a computer for modeling a newly-constructed nuclear reactor

   D. Use of a pen-and-paper approach to a difficult problem

PHYSICS

4. Which of the following is not a key purpose for the use of open communication about and peer-review of the results of scientific investigations?
(Average Rigor)

   A. Testing, by other scientists, of the results of an investigation for the purpose of refuting any evidence contrary to an established theory

   B. Testing, by other scientists, of the results of an investigation for the purpose of finding or eliminating any errors in reasoning or measurement

   C. Maintaining an open, public process to better promote honesty and integrity in science

   D. Provide a forum to help promote progress through mutual sharing and review of the results of investigations

5. Which of the following aspects of the use of computers for collecting experimental data is not a concern for the scientist?
(Rigorous)

   A. The relative speeds of the processor, peripheral, memory storage unit and any other components included in data acquisition equipment

   B. The financial cost of the equipment, utilities and maintenance

   C. Numerical error due to a lack of infinite precision in digital equipment

   D. The order of complexity of data analysis algorithms

6. **If a particular experimental observation contradicts a theory, what is the most appropriate approach that a physicist should take?**
   (Average Rigor)

   A. Immediately reject the theory and begin developing a new theory that better fits the observed results

   B. Report the experimental result in the literature without further ado

   C. Repeat the observations and check the experimental apparatus for any potential faulty components or human error, and then compare the results once more with the theory

   D. Immediately reject the observation as in error due to its conflict with theory

7. **Which of the following is *not* an SI unit?**
   (Average Rigor)

   A. Joule

   B. Coulomb

   C. Newton

   D. Erg

8. **Which of the following best describes the relationship of precision and accuracy in scientific measurements?**
   (Easy)

   A. Accuracy is how well a particular measurement agrees with the value of the actual parameter being measured; precision is how well a particular measurement agrees with the average of other measurements taken for the same value

   B. Precision is how well a particular measurement agrees with the value of the actual parameter being measured; accuracy is how well a particular measurement agrees with the average of other measurements taken for the same value

   C. Accuracy is the same as precision

   D. Accuracy is a measure of numerical error; precision is a measure of human error

9. Which statement best describes a rationale for the use of statistical analysis in characterizing the numerical results of a scientific experiment or investigation?
(*Average Rigor*)

   A. Experimental results need to be adjusted, through the use of statistics, to conform to theoretical predictions and computer models

   B. Since experiments are prone to a number of errors and uncertainties, statistical analysis provides a method for characterizing experimental measurements by accounting for or quantifying these undesirable effects

   C. Experiments are not able to provide any useful information, and statistical analysis is needed to impose a theoretical framework on the results

   D. Statistical analysis is needed to relate experimental measurements to computer-simulated values

10. Which statement best characterizes the relationship of mathematics and experimentation in physics?
(*Easy*)

   A. Experimentation has no bearing on the mathematical models that are developed for physical phenomena

   B. Mathematics is a tool that assists in the development of models for various physical phenomena as they are studied experimentally, with observations of the phenomena being a test of the validity of the mathematical model

   C. Mathematics is used to test the validity of experimental apparatus for physical measurements

   D. Mathematics is an abstract field with no relationship to concrete experimentation

11. Which of the following mathematical tools would not typically be used for the analysis of an electromagnetic phenomenon?
    (Rigorous)

    A. Trigonometry

    B. Vector calculus

    C. Group theory

    D. Numerical methods

12. For a problem that involves parameters that vary in rate with direction and location, which of the following mathematical tools would most likely be of greatest value?
    (Rigorous)

    A. Trigonometry

    B. Numerical analysis

    C. Group theory

    D. Vector calculus

13. Which of the following devices would be best suited for an experiment designed to measure alpha particle emissions from a sample?
    (Average Rigor)

    A. Photomultiplier tube

    B. Thermocouple

    C. Geiger-Müller tube

    D. Transistor

14. Which of the following experiments presents the most likely cause for concern about laboratory safety?
    (Average Rigor)

    A. Computer simulation of a nuclear reactor

    B. Vibration measurement with a laser

    C. Measurement of fluorescent light intensity with a battery-powered photodiode circuit

    D. Ambient indoor ionizing radiation measurement with a Geiger counter.

15. A brick and hammer fall from a ledge at the same time. They would be expected to:
    (Easy)

    A. Reach the ground at the same time

    B. Accelerate at different rates due to difference in weight

    C. Accelerate at different rates due to difference in potential energy

    D. Accelerate at different rates due to difference in kinetic energy

16. A baseball is thrown with an initial velocity of 30 m/s at an angle of 45°. Neglecting air resistance, how far away will the ball land?
    (Rigorous)

    A. 92 m

    B. 78 m

    C. 65 m

    D. 46 m

17. A skateboarder accelerates down a ramp, with constant acceleration of two meters per second squared, from rest. The distance in meters, covered after four seconds, is:
    (Rigorous)

    A. 10

    B. 16

    C. 23

    D. 37

18. When acceleration is plotted versus time, the area under the graph represents:
    (Average Rigor)

    A. Time

    B. Distance

    C. Velocity

    D. Acceleration

19. An inclined plane is tilted by gradually increasing the angle of elevation θ, until the block will slide down at a constant velocity. The coefficient of friction, μ_k, is given by:
(Rigorous)

A. cos θ

B. sin θ

C. cosecant θ

D. tangent θ

20. An object traveling through air loses part of its energy of motion due to friction. Which statement best describes what has happened to this energy?
(Easy)

A. The energy is destroyed

B. The energy is converted to static charge

C. The energy is radiated as electromagnetic waves

D. The energy is lost to heating of the air

21. The weight of an object on the earth's surface is designated x. When it is two earth's radii from the surface of the earth, its weight will be:
(Rigorous)

A. $x/4$

B. $x/9$

C. $4x$

D. $16x$

22. Which of the following units is not used to measure torque?
(Average Rigor)

A. slug ft

B. lb ft

C. N m

D. dyne cm

23. A uniform pole weighing 100 grams, that is one meter in length, is supported by a pivot at 40 centimeters from the left end. In order to maintain static position, a 200 gram mass must be placed _____ centimeters from the left end.
(Rigorous)

   A. 10

   B. 45

   C. 35

   D. 50

24. The magnitude of a force is:
(Easy)

   A. Directly proportional to mass and inversely to acceleration

   B. Inversely proportional to mass and directly to acceleration

   C. Directly proportional to both mass and acceleration

   D. Inversely proportional to both mass and acceleration

25. A projectile with a mass of 1.0 kg has a muzzle velocity of 1500.0 m/s when it is fired from a cannon with a mass of 500.0 kg. If the cannon slides on a frictionless track, it will recoil with a velocity of ____ m/s.
(Rigorous)

   A. 2.4

   B. 3.0

   C. 3.5

   D. 1500

26. A car (mass m_1) is driving at velocity v, when it smashes into an unmoving car (mass m_2), locking bumpers. Both cars move together at the same velocity. The common velocity will be given by:
(Rigorous)

   A. m_1v/m_2

   B. m_2v/m_1

   C. $m_1v/(m_1 + m_2)$

   D. $(m_1 + m_2)v/m_1$

PHYSICS

27. A satellite is in a circular orbit above the earth. Which statement is false?
    (Average Rigor)

    A. An external force causes the satellite to maintain orbit.

    B. The satellite's inertia causes it to maintain orbit.

    C. The satellite is accelerating toward the earth.

    D. The satellite's velocity and acceleration are not in the same direction.

28. A 100 g mass revolving around a fixed point, on the end of a 0.5 meter string, circles once every 0.25 seconds. What is the magnitude of the centripetal acceleration?
    (Average Rigor)

    A. 1.23 m/s^2

    B. 31.6 m/s^2

    C. 100 m/s^2

    D. 316 m/s^2

29. Which statement best describes the relationship of simple harmonic motion to a simple pendulum of length L, mass m and displacement of arc length s?
    (Average Rigor)

    A. A simple pendulum cannot be modeled using simple harmonic motion

    B. A simple pendulum may be modeled using the same expression as Hooke's law for displacement s, but with a spring constant equal to the tension on the string

    C. A simple pendulum may be modeled using the same expression as Hooke's law but with a spring constant equal to m g/L

    D. A simple pendulum typically does not undergo simple harmonic motion

30. A mass of 2 kg connected to a spring undergoes simple harmonic motion at a frequency of 3 Hz. What is the spring constant?
    (Average Rigor)

    A. 6 kg/s^2

    B. 18 kg/s^2

    C. 710 kg/s^2

    D. 1000 kg/s^2

31. The kinetic energy of an object is _____ proportional to its _____.
(Average Rigor)

   A. Inversely…inertia

   B. Inversely…velocity

   C. Directly…mass

   D. Directly…time

32. A force is given by the vector 5 N x + 3 N y (where x and y are the unit vectors for the x- and y- axes, respectively). This force is applied to move a 10 kg object 5 m, in the x direction. How much work was done?
(Rigorous)

   A. 250 J

   B. 400 J

   C. 40 J

   D. 25 J

33. An office building entry ramp uses the principle of which simple machine?
(Easy)

   A. Lever

   B. Pulley

   C. Wedge

   D. Inclined Plane

34. If the internal energy of a system remains constant, how much work is done by the system if 1 kJ of heat energy is added?
(Average Rigor)

   A. 0 kJ

   B. -1 kJ

   C. 1 kJ

   D. 3.14 kJ

35. A calorie is the amount of heat energy that will: *(Easy)*

    A. Raise the temperature of one gram of water from 14.5° C to 15.5° C.

    B. Lower the temperature of one gram of water from 16.5° C to 15.5° C

    C. Raise the temperature of one gram of water from 32° F to 33° F

    D. Cause water to boil at two atmospheres of pressure.

36. An ice block at 0° Celsius is dropped into 100 g of liquid water at 18° Celsius. When thermal equilibrium is achieved, only liquid water at 0° Celsius is left. What was the mass, in grams, of the original block of ice?
    Given:
    1. Heat of fusion of ice = 80 cal/g
    2. Heat of vaporization of ice = 540 cal/g
    3. Specific Heat of ice = 0.50 cal/g°C
    4. Specific Heat of water = 1 cal/g°C

    (Rigorous)

    A. 2.0

    B. 5.0

    C. 10.0

    D. 22.5

37. Heat transfer by electromagnetic waves is termed: *(Easy)*

    A. Conduction

    B. Convection

    C. Radiation

    D. Phase Change

38. A cooking thermometer in an oven works because the metals it is composed of have different: *(Average Rigor)*

    A. Melting points

    B. Heat convection

    C. Magnetic fields

    D. Coefficients of expansion

39. Which of the following is not an assumption upon which the kinetic-molecular theory of gases is based?
    (Rigorous)

    A. Quantum mechanical effects may be neglected

    B. The particles of a gas may be treated statistically

    C. The particles of the gas are treated as very small masses

    D. Collisions between gas particles and container walls are inelastic

40. What is temperature?
    (Average Rigor)

    A. Temperature is a measure of the conductivity of the atoms or molecules in a material

    B. Temperature is a measure of the kinetic energy of the atoms or molecules in a material

    C. Temperature is a measure of the relativistic mass of the atoms or molecules in a material

    D. Temperature is a measure of the angular momentum of electrons in a material

41. Solids expand when heated because:
    (Rigorous)

    A. Molecular motion causes expansion

    B. PV = nRT

    C. Magnetic forces stretch the chemical bonds

    D. All material is effectively fluid

42. What should be the behavior of an electroscope, which has been grounded in the presence of a positively charged object (1), after the ground connection is removed and then the charged object is removed from the vicinity (2)? *(Average Rigor)*

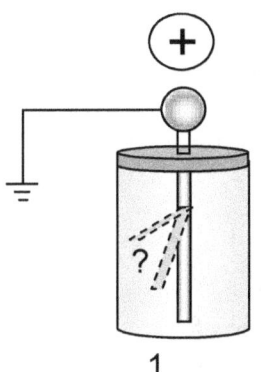
1

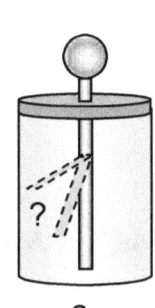
2

A. The metal leaf will start deflected (1) and then relax to an undeflected position (2)

B. The metal leaf will start in an undeflected position (1) and then be deflected (2)

C. The metal leaf will remain undeflected in both cases

D. The metal leaf will be deflected in both cases

43. The electric force in Newtons, on two small objects (each charged to – 10 microCoulombs and separated by 2 meters) is: *(Rigorous)*

A. 1.0

B. 9.81

C. 31.0

D. 0.225

44. A 10 ohm resistor and a 50 ohm resistor are connected in parallel. If the current in the 10 ohm resistor is 5 amperes, the current (in amperes) running through the 50 ohm resistor is: *(Rigorous)*

A. 1

B. 50

C. 25

D. 60

PHYSICS

45. How much power is dissipated through the following resistive circuit?
(Average Rigor)

A. 0 W

B. 0.22 W

C. 0.31 W

D. 0.49 W

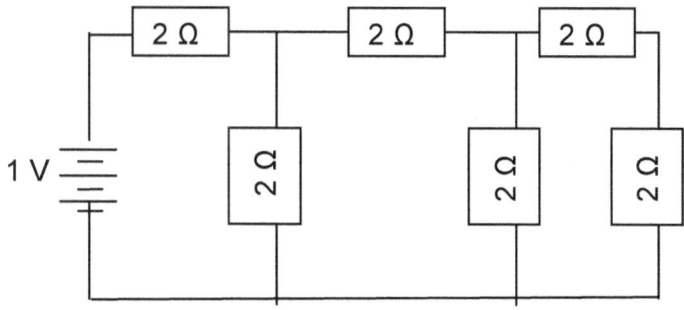

46. The greatest number of 100 watt lamps that can be connected in parallel with a 120 volt system without blowing a 5 amp fuse is:
(Rigorous)

A. 24

B. 12

C. 6

D. 1

47. Which of the following statements may be taken as a legitimate inference based upon the Maxwell equation that states $\nabla \cdot \mathbf{B} = 0$?
(Average Rigor)

A. The electric and magnetic fields are decoupled

B. The electric and magnetic fields are mediated by the W boson

C. There are no photons

D. There are no magnetic monopoles

48. What effect might an applied external magnetic field have on the magnetic domains of a ferromagnetic material?
(Rigorous)

A. The domains that are not aligned with the external field increase in size, but those that are aligned decrease in size

B. The domains that are not aligned with the external field decrease in size, but those that are aligned increase in size

C. The domains align perpendicular to the external field

D. There is no effect on the magnetic domains

49. What is the effect of running current in the same direction along two parallel wires, as shown below? *(Rigorous)*

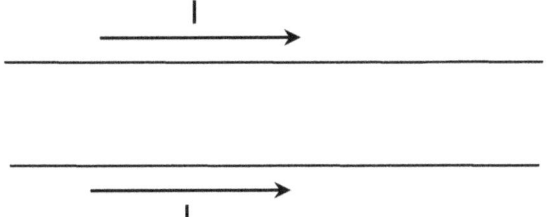

   A. There is no effect

   B. The wires attract one another

   C. The wires repel one another

   D. A torque is applied to both wires

50. The current induced in a coil is defined by which of the following laws? *(Easy)*

   A. Lenz's Law

   B. Burke's Law

   C. The Law of Spontaneous Combustion

   D. Snell's Law

51. A light bulb is connected in series with a rotating coil within a magnetic field. The brightness of the light may be increased by any of the following except: *(Average Rigor)*

   A. Rotating the coil more rapidly.

   B. Using more loops in the coil.

   C. Using a different color wire for the coil.

   D. Using a stronger magnetic field.

52. What is the direction of the magnetic field at the center of the loop of current (I) shown below (i.e., at point A)? *(Easy)*

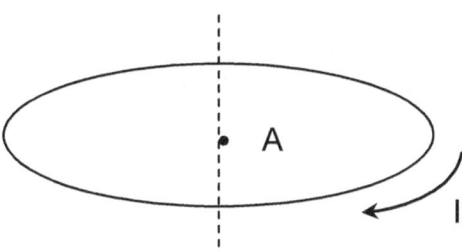

   A. Down, along the axis (dotted line)

   B. Up, along the axis (dotted line)

   C. The magnetic field is oriented in a radial direction

   D. There is no magnetic field at point A

53. The use of two circuits next to each other, with a change in current in the primary circuit, demonstrates:
    (Rigorous)

    A. Mutual current induction

    B. Dielectric constancy

    C. Harmonic resonance

    D. Resistance variation

54. A semi-conductor allows current to flow:
    (Easy)

    A. Never

    B. Always

    C. As long as it stays below a maximum temperature

    D. When a minimum voltage is applied

55. All of the following use semi-conductor technology, except a(n):
    (Average Rigor)

    A. Transistor

    B. Diode

    C. Capacitor

    D. Operational Amplifier

56. A wave generator is used to create a succession of waves. The rate of wave generation is one every 0.33 seconds. The period of these waves is:
    (Average Rigor)

    A. 2.0 seconds

    B. 1.0 seconds

    C. 0.33 seconds

    D. 3.0 seconds

57. An electromagnetic wave propagates through a vacuum. Independent of its wavelength, it will move with constant:
    (Easy)

    A. Acceleration

    B. Velocity

    C. Induction

    D. Sound

58. A wave has speed 60 m/s and wavelength 30,000 m. What is the frequency of the wave?
    (Average Rigor)

    A. 2.0×10^{-3} Hz

    B. 60 Hz

    C. 5.0×10^2 Hz

    D. 1.8×10^6 Hz

59. Rainbows are created by:
    (Easy)

    A. Reflection, dispersion, and recombination

    B. Reflection, resistance, and expansion

    C. Reflection, compression, and specific heat

    D. Reflection, refraction, and dispersion

60. Which of the following is *not* a legitimate explanation for refraction of light rays at boundaries between different media?
    (Rigorous)

    A. Light seeks the path of least time between two different points

    B. Due to phase matching and other boundary conditions, plane waves travel in different directions on either side of the boundary, depending on the material parameters

    C. The electric and magnetic fields become decoupled at the boundary

    D. Light rays obey Snell's law

61. A stationary sound source produces a wave of frequency *F*. An observer at position A is moving toward the horn, while an observer at position B is moving away from the horn. Which of the following is true?
(Rigorous)

   A. $F_A < F < F_B$

   B. $F_B < F < F_A$

   C. $F < F_A < F_B$

   D. $F_B < F_A < F$

62. A monochromatic ray of light passes from air to a thick slab of glass (n = 1.41) at an angle of 45° from the normal. At what angle does it leave the air/glass interface?
(Rigorous)

   A. 45°

   B. 30°

   C. 15°

   D. 55°

63. If one sound is ten decibels louder than another, the ratio of the intensity of the first to the second is:
(Average Rigor)

   A. 20:1

   B. 10:1

   C. 1:1

   D. 1:10

64. The velocity of sound is greatest in:
(Average Rigor)

   A. Water

   B. Steel

   C. Alcohol

   D. Air

65. A vibrating string's frequency is _____ proportional to the _____.
    (Rigorous)

    A. Directly; Square root of the tension

    B. Inversely; Length of the string

    C. Inversely; Squared length of the string

    D. Inversely; Force of the plectrum

66. Which of the following apparatus can be used to measure the wavelength of a sound produced by a tuning fork?
    (Average Rigor)

    A. A glass cylinder, some water, and iron filings

    B. A glass cylinder, a meter stick, and some water

    C. A metronome and some ice water

    D. A comb and some tissue

67. The highest energy is associated with:
    (Easy)

    A. UV radiation

    B. Yellow light

    C. Infrared radiation

    D. Gamma radiation

68. An object two meters tall is speeding toward a plane mirror at 10 m/s. What happens to the image as it nears the surface of the mirror?
    (Rigorous)

    A. It becomes inverted.

    B. The Doppler Effect must be considered.

    C. It remains two meters tall.

    D. It changes from a real image to a virtual image.

69. **Automobile mirrors that have a sign, "objects are closer than they appear" say so because:**
(Rigorous)

   A. The real image of an obstacle, through a converging lens, appears farther away than the object.

   B. The real or virtual image of an obstacle, through a converging mirror, appears farther away than the object.

   C. The real image of an obstacle, through a diverging lens, appears farther away than the object.

   D. The virtual image of an obstacle, through a diverging mirror, appears farther away than the object.

70. **If an object is 20 cm from a convex lens whose focal length is 10 cm, the image is:**
(Rigorous)

   A. Virtual and upright

   B. Real and inverted

   C. Larger than the object

   D. Smaller than the object

71. **The constant of proportionality between the energy and the frequency of electromagnetic radiation is known as the:**
(Easy)

   A. Rydberg constant

   B. Energy constant

   C. Planck constant

   D. Einstein constant

72. **Which phenomenon was first explained using the concept of quantization of energy, thus providing one of the key foundational principles for the later development of quantum theory?**
(Rigorous)

   A. The photoelectric effect

   B. Time dilation

   C. Blackbody radiation

   D. Magnetism

73. **Which statement best describes why population inversion is necessary for a laser to operate?**
    (Rigorous)

    A. Population inversion prevents too many electrons from being excited into higher energy levels, thus preventing damage to the gain medium.

    B. Population inversion maintains a sufficient number of electrons in a higher energy state so as to allow a significant amount of stimulated emission.

    C. Population inversion prevents the laser from producing coherent light.

    D. Population inversion is not necessary for the operation of most lasers.

74. **Bohr's theory of the atom was the first to quantize:**
    (Average Rigor)

    A. Work

    B. Angular Momentum

    C. Torque

    D. Duality

75. **Two neutral isotopes of a chemical element have the same numbers of:**
    (Easy)

    A. Electrons and Neutrons

    B. Electrons and Protons

    C. Protons and Neutrons

    D. Electrons, Neutrons, and Protons

76. **When a radioactive material emits an alpha particle only, its atomic number will:**
    (Average Rigor)

    A. Decrease

    B. Increase

    C. Remain unchanged

    D. Change randomly

77. **Ten grams of a sample of a radioactive material (half-life = 12 days) were stored for 48 days and re-weighed. The new mass of material was:**
    (Rigorous)

    A. 1.25 g

    B. 2.5 g

    C. 0.83 g

    D. 0.625 g

78. Which of the following pairs of elements are not found to fuse in the centers of stars?
   (Average Rigor)

   A. Oxygen and Helium

   B. Carbon and Hydrogen

   C. Beryllium and Helium

   D. Cobalt and Hydrogen

79. In a fission reactor, heavy water:
   (Average Rigor)

   A. Cools off neutrons to control temperature

   B. Moderates fission reactions

   C. Initiates the reaction chain

   D. Dissolves control rods

80. Given the following values for the masses of a proton, a neutron and an alpha particle, what is the nuclear binding energy of an alpha particle?
   (Rigorous)

   Proton mass=1.6726 x 10^{-27} kg
   Neutron mass=1.6749 x 10^{-27} kg
   Alpha particle mass= 6.6465 x 10^{-27} kg

   A. 0 J

   B. 7.3417 x 10^{-27} J

   C. 4 J

   D. 4.3589 x 10^{-12} J

Answer Key

1.	D	25.	B	49.	B	73.	B
2.	A	26.	C	50.	A	74.	B
3.	B	27.	B	51.	C	75.	B
4.	A	28.	D	52.	A	76.	A
5.	D	29.	C	53.	A	77.	D
6.	C	30.	C	54.	D	78.	D
7.	D	31.	C	55.	C	79.	B
8.	A	32.	D	56.	C	80.	D
9.	B	33.	D	57.	B		
10.	B	34.	C	58.	A		
11.	C	35.	A	59.	D		
12.	D	36.	D	60.	C		
13.	C	37.	C	61.	B		
14.	B	38.	D	62.	B		
15.	A	39.	D	63.	B		
16.	A	40.	B	64.	B		
17.	B	41.	A	65.	A		
18.	C	42.	B	66.	B		
19.	D	43.	D	67.	D		
20.	D	44.	A	68.	C		
21.	B	45.	C	69.	D		
22.	A	46.	C	70.	B		
23.	C	47.	D	71.	C		
24.	C	48.	B	72.	C		

Rigor Analysis Table

Easy 21%	1,8,10,15,20,24,33,35,37,50,52,54,57,59,67,71,75
Average Rigor 39%	2,4,6,7,9,13,14,18,22,27,28,29,30,31,34,38,40,42, 45,47,51,55,56,58,63,64,66,74,76,78,79
Rigorous 40%	3,5,11,12,16,17,19,21,23,25,26,32,36,39,41,43,44, 46,48,49,53,60,61,62,65,68,69,70,72,73,77,80

TEACHER CERTIFICATION STUDY GUIDE

Rationales with Sample Questions

1. **Which statement best describes a valid approach to testing a scientific hypothesis?**
   (Easy)

      A. Use computer simulations to verify the hypothesis

      B. Perform a mathematical analysis of the hypothesis

      C. Design experiments to test the hypothesis

      D. All of the above

Answer: D

Each of the answers A, B and C can have a crucial part in testing a scientific hypothesis. Although experiments may hold more weight than mathematical or computer-based analysis, these latter two methods of analysis can be critical, especially when experimental design is highly time consuming or financially costly.

2. **Which description best describes the role of a scientific model of a physical phenomenon?**
   (Average Rigor)

      A. An explanation that provides a reasonably accurate approximation of the phenomenon

      B. A theoretical explanation that describes exactly what is taking place

      C. A purely mathematical formulation of the phenomenon

      D. A predictive tool that has no interest in what is actually occurring

Answer: A

A scientific model seeks to provide the most fundamental and accurate description possible for physical phenomena, but, given the fact that natural science takes an *a posteriori* approach, models are always tentative and must be treated with some amount of skepticism. As a result, A is a better answer than B. Answers C and D overly emphasize one or another aspect of a model, rather than a synthesis of a number of aspects (such as a mathematical and predictive aspect).

3. **Which situation calls might best be described as involving an ethical dilemma for a scientist?**
(Rigorous)

    A. Submission to a peer-review journal of a paper that refutes an established theory

    B. Synthesis of a new radioactive isotope of an element

    C. Use of a computer for modeling a newly-constructed nuclear reactor

    D. Use of a pen-and-paper approach to a difficult problem

Answer: B

Although answer A may be controversial, it does not involve an inherently ethical dilemma, since there is nothing unethical about presenting new information if it is true or valid. Answer C, likewise, has no necessary ethical dimension, as is the case with D. Synthesis of radioactive material, however, involves an ethical dimension with regard to the potential impact of the new isotope on the health of others and on the environment. The potential usefulness of such an isotope in weapons development is another ethical consideration.

4. **Which of the following is not a key purpose for the use of open communication about and peer-review of the results of scientific investigations?**
(Average Rigor)

    A. Testing, by other scientists, of the results of an investigation for the purpose of refuting any evidence contrary to an established theory

    B. Testing, by other scientists, of the results of an investigation for the purpose of finding or eliminating any errors in reasoning or measurement

    C. Maintaining an open, public process to better promote honesty and integrity in science

    D. Provide a forum to help promote progress through mutual sharing and review of the results of investigations

Answer: A

Answers B, C and D all are important rationales for the use of open communication and peer-review in science. Answer A, however, would suggest that the purpose of these processes is to simply maintain the status quo; the history of science, however, suggests that this cannot and should not be the case.

5. **Which of the following aspects of the use of computers for collecting experimental data is not a concern for the scientist?**
   (Rigorous)

   A. The relative speeds of the processor, peripheral, memory storage unit and any other components included in data acquisition equipment

   B. The financial cost of the equipment, utilities and maintenance

   C. Numerical error due to a lack of infinite precision in digital equipment

   D. The order of complexity of data analysis algorithms

Answer: D

Although answer D might be a concern for later, when actual analysis of the data is undertaken, the collection of data typically does not suffer from this problem. The use of computers does, however, pose problems when, for example, a peripheral collects data at a rate faster than the computer can process it (A), or if the cost of running the equipment or of purchasing the equipment is prohibitive (B). Numerical error is always a concern with any digital data acquisition system, since the data that is collected is never exact.

6. **If a particular experimental observation contradicts a theory, what is the most appropriate approach that a physicist should take?**
(Average Rigor)

    A. Immediately reject the theory and begin developing a new theory that better fits the observed results

    B. Report the experimental result in the literature without further ado

    C. Repeat the observations and check the experimental apparatus for any potential faulty components or human error, and then compare the results once more with the theory

    D. Immediately reject the observation as in error due to its conflict with theory

Answer: C

When experimental results contradict a reigning physical theory, as they do from time to time, it is almost never appropriate to immediately reject the theory (A) *or* the observational results (D). Also, since this is the case, reporting the result in the literature, without further analysis to provide an adequate explanation of the discrepancy, is unwise and unwarranted. Further testing is appropriate to determine whether the experiment is repeatable and whether any equipment or human errors have occurred. Only after further testing may the physicist begin to analyze the implications of the observational result.

7. **Which of the following is *not* an SI unit?**
(Average Rigor)

    A. Joule

    B. Coulomb

    C. Newton

    D. Erg

Answer: D

The first three responses are the SI (*Le Système International d'Unités*) units for energy, charge and force, respectively. The fourth answer, the erg, is the CGS (centimeter-gram-second) unit of energy.

TEACHER CERTIFICATION STUDY GUIDE

8. **Which of the following best describes the relationship of precision and accuracy in scientific measurements?**
   (Easy)

   A. Accuracy is how well a particular measurement agrees with the value of the actual parameter being measured; precision is how well a particular measurement agrees with the average of other measurements taken for the same value

   B. Precision is how well a particular measurement agrees with the value of the actual parameter being measured; accuracy is how well a particular measurement agrees with the average of other measurements taken for the same value

   C. Accuracy is the same as precision

   D. Accuracy is a measure of numerical error; precision is a measure of human error

Answer: A

The accuracy of a measurement is how close the measurement is to the "true" value of the parameter being measured. Precision is how closely a group of measurements is to the mean value of all the measurements. By analogy, accuracy is how close a measurement is to the center of the bulls-eye, and precision is how tight a group is formed by multiple measurements, regardless of accuracy. Thus, measurements may be very precise and not very accurate, or they may be accurate but not overly precise, or they may be both or neither.

9. **Which statement best describes a rationale for the use of statistical analysis in characterizing the numerical results of a scientific experiment or investigation?**
(Average Rigor)

   A. Experimental results need to be adjusted, through the use of statistics, to conform to theoretical predictions and computer models

   B. Since experiments are prone to a number of errors and uncertainties, statistical analysis provides a method for characterizing experimental measurements by accounting for or quantifying these undesirable effects

   C. Experiments are not able to provide any useful information, and statistical analysis is needed to impose a theoretical framework on the results

   D. Statistical analysis is needed to relate experimental measurements to computer-simulated values

Answer: B

One of the main reasons for the use of statistical analysis is that various types of noise, errors and uncertainties can easily enter into experimental results. Among other things, statistics can help alleviate these difficulties by quantifying an average measurement value and a variance or standard deviation of the set of measurements. This helps determine the accuracy and precision of a set of experimental results. Answers A, C and D do not accurately describe ideal scientific experiments or the use of statistics.

10. **Which statement best characterizes the relationship of mathematics and experimentation in physics?**
    (Easy)

    A. Experimentation has no bearing on the mathematical models that are developed for physical phenomena

    B. Mathematics is a tool that assists in the development of models for various physical phenomena as they are studied experimentally, with observations of the phenomena being a test of the validity of the mathematical model

    C. Mathematics is used to test the validity of experimental apparatus for physical measurements

    D. Mathematics is an abstract field with no relationship to concrete experimentation

Answer: B

Mathematics is used extensively in the study of physics for creating models of various phenomena. Since mathematics is abstract and not necessarily tied to physical reality, it must be tempered by experimental results. Although a particular theory may be mathematically elegant, it may have no explanatory power due to its inability to account for certain aspects of physical reality, or due to its inclusion of gratuitous aspects that seem to have no physical analog. Thus, experimentation is foundational, with mathematics being a tool for organizing and providing a greater context for observational results.

11. **Which of the following mathematical tools would not typically be used for the analysis of an electromagnetic phenomenon?**
    (Rigorous)

    A. Trigonometry

    B. Vector calculus

    C. Group theory

    D. Numerical methods

Answer: C

Trigonometry and vector calculus are both key tools for solving problems in electromagnetics. These are, primarily, analytical methods, although they play a part in numerical analysis as well. Numerical methods are helpful for many problems that are otherwise intractable analytically. Group theory, although it may have some applications in certain highly specific areas, is generally not used in the study of electromagnetics.

TEACHER CERTIFICATION STUDY GUIDE

12. For a problem that involves parameters that vary in rate with direction and location, which of the following mathematical tools would most likely be of greatest value?
    (Rigorous)

    A. Trigonometry

    B. Numerical analysis

    C. Group theory

    D. Vector calculus

Answer: D

Each of the above answers might have some value for individual problems, but, generally speaking, those problems that deal with quantities that have direction and magnitude (vectors), and that deal with rates, would most likely be amenable to analysis using vector calculus (D).

13. Which of the following devices would be best suited for an experiment designed to measure alpha particle emissions from a sample?
    (Average Rigor)

    A. Photomultiplier tube

    B. Thermocouple

    C. Geiger-Müller tube

    D. Transistor

Answer: C

The Geiger-Müller tube is the main component of the so-called Geiger counter, which is designed specifically for detecting ionizing radiation emissions, including alpha particles. The photomultiplier tube is better suited to measurement of electromagnetic radiation closer to the visible range (A), and the thermocouple is better suited to measurement of temperature (B). Transistors may be involved in instrumentation, but they are not sensors.

TEACHER CERTIFICATION STUDY GUIDE

14. **Which of the following experiments presents the most likely cause for concern about laboratory safety?**
    (Average Rigor)

    A. Computer simulation of a nuclear reactor

    B. Vibration measurement with a laser

    C. Measurement of fluorescent light intensity with a battery-powered photodiode circuit

    D. Ambient indoor ionizing radiation measurement with a Geiger counter.

Answer: B

Assuming no profoundly foolish acts, the use of a computer for simulation (A), measurement with a battery-powered photodiode circuit (C) and ambient radiation measurement (D) pose no particular hazards. The use of a laser (B) must be approached with care, however, as unintentional reflections or a lack of sufficient protection can cause permanent eye damage.

15. **A brick and hammer fall from a ledge at the same time. They would be expected to:**
    (Easy)

    A. Reach the ground at the same time

    B. Accelerate at different rates due to difference in weight

    C. Accelerate at different rates due to difference in potential energy

    D. Accelerate at different rates due to difference in kinetic energy

Answer: A

This is a classic question about falling in a gravitational field. All objects are acted upon equally by gravity, so they should reach the ground at the same time. (In real life, air resistance can make a difference, but not at small heights for similarly shaped objects.) In any case, weight, potential energy, and kinetic energy do not affect gravitational acceleration. Thus, the only possible answer is (A).

TEACHER CERTIFICATION STUDY GUIDE

16. A baseball is thrown with an initial velocity of 30 m/s at an angle of 45°. Neglecting air resistance, how far away will the ball land?
(Rigorous)

    A. 92 m

    B. 78 m

    C. 65 m

    D. 46 m

Answer: A

To answer this question, recall the equations for projectile motion:
$y = \frac{1}{2} a t^2 + v_{0y} t + y_0$
$x = v_{0x} t + x_0$
where x and y are horizontal and vertical position, respectively; t is time; a is acceleration due to gravity; v_{0x} and v_{0y} are initial horizontal and vertical velocity, respectively; x_0 and y_0 are initial horizontal and vertical position, respectively.
For our case:
x_0 and y_0 can be set to zero
both v_{0x} and v_{0y} are (using trigonometry) = $(\sqrt{2}/2)$ 30 m/s
$a = -9.81$ m/s^2

We then use the vertical motion equation to find the time aloft (setting y equal to zero to find the solution for t):
$0 = \frac{1}{2} (-9.81 \text{ m/s}^2) t^2 + (\sqrt{2}/2)$ 30 m/s t
Then solving, we find:
t = 0 s (initial set-up) or t = 4.324 s (time to go up and down)

Using t = 4.324 s in the horizontal motion equation, we find:
$x = ((\sqrt{2}/2)$ 30 m/s) (4.324 s)
x = 91.71 m

This is consistent only with answer (A).

TEACHER CERTIFICATION STUDY GUIDE

17. A skateboarder accelerates down a ramp, with constant acceleration of two meters per second squared, from rest. The distance in meters, covered after four seconds, is:
    (Rigorous)

    A. 10

    B. 16

    C. 23

    D. 37

Answer: B

To answer this question, recall the equation relating constant acceleration to distance and time:
$x = \frac{1}{2} a t^2 + v_0 t + x_0$ where x is position; a is acceleration; t is time; v_0 and x_0 are initial velocity and position (both zero in this case)

thus, to solve for x:
$x = \frac{1}{2} (2 \text{ m/s}^2)(4^2 \text{s}^2) + 0 + 0$
$x = 16 \text{ m}$

This is consistent only with answer (B).

18. When acceleration is plotted versus time, the area under the graph represents:
    (Average Rigor)

    A. Time

    B. Distance

    C. Velocity

    D. Acceleration

Answer: C
The area under a graph will have units equal to the product of the units of the two axes. (To visualize this, picture a graphed rectangle with its area equal to length times width.)
Therefore, multiply units of acceleration by units of time:
$(\text{length/time}^2)(\text{time})$
This equals length/time, i.e. units of velocity.

PHYSICS

19. An inclined plane is tilted by gradually increasing the angle of elevation θ, until the block will slide down at a constant velocity. The coefficient of friction, μ$_k$, is given by:
    (Rigorous)

    A. cos θ

    B. sin θ

    C. cosecant θ

    D. tangent θ

Answer: D

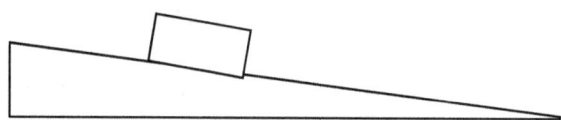

When the block moves, its force upstream (due to friction) must equal its force downstream (due to gravity).

The friction force is given by
$F_f = \mu_k N$
where μ$_k$ is the friction coefficient and N is the normal force.

Using similar triangles, the gravity force is given by
$F_g = mg \sin \theta$
and the normal force is given by
$N = mg \cos \theta$

When the block moves at constant velocity, it must have zero net force, so set equal the force of gravity and the force due to friction:
$F_f = F_g$
$\mu_k mg \cos \theta = mg \sin \theta$
$\mu_k = \tan \theta$

Answer (D) is the only appropriate choice in this case.

20. An object traveling through air loses part of its energy of motion due to friction. Which statement best describes what has happened to this energy?
(Easy)

    A. The energy is destroyed

    B. The energy is converted to static charge

    C. The energy is radiated as electromagnetic waves

    D. The energy is lost to heating of the air

Answer: D

Since energy must be conserved, the energy of motion of the object is converted, in part, to energy of motion of the molecules in the air (and, to some extent, in the object). This additional motion is equivalent to an increase in heat. Thus, friction is a loss of energy of motion through heating.

21. The weight of an object on the earth's surface is designated x. When it is two earth's radii from the surface of the earth, its weight will be:
(Rigorous)

    A. $x/4$

    B. $x/9$

    C. $4x$

    D. $16x$

Answer: B

To solve this problem, apply the universal Law of Gravitation to the object and Earth:

$F_{gravity} = (GM_1M_2)/R^2$

Because the force of gravity varies with the square of the radius between the objects, the force (or weight) on the object will be decreased by the square of the multiplication factor on the radius. Note that the object on Earth's surface is *already* at one radius from Earth's center. Thus, when it is two radii from Earth's surface, it is three radii from Earth's center. R^2 is then nine, so the weight is $x/9$. Only answer (B) matches these calculations.

22. Which of the following units is not used to measure torque? *(Average Rigor)*

    A. slug ft

    B. lb ft

    C. N m

    D. dyne cm

Answer: A

To answer this question, recall that torque is always calculated by multiplying units of force by units of distance. Therefore, answer (A), which is the product of units of mass and units of distance, must be the choice of incorrect units. Indeed, the other three answers all could measure torque, since they are of the correct form. It is a good idea to review "English Units" before the teacher test, because they are occasionally used in problems.

23. A uniform pole weighing 100 grams, that is one meter in length, is supported by a pivot at 40 centimeters from the left end. In order to maintain static position, a 200 gram mass must be placed _____ centimeters from the left end.
(Rigorous)

    A. 10

    B. 45

    C. 35

    D. 50

Answer: C

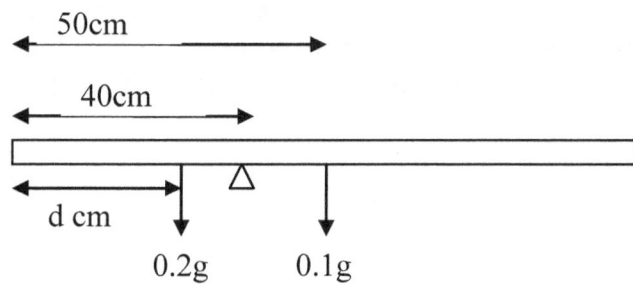

Since the pole is uniform, we can assume that its weight 0.1g acts at the center, i.e. 50 cm from the left end. In order to keep the pole balanced on the pivot, the 200 gram mass must be placed such that the torque on the pole due to the mass is equal and opposite to the torque due to the pole's weight. Thus, if the 200 gram mass is placed d cm from the left end of the pole,

$(40 - d) \times 0.2g = 10 \times 0.1g$; $40 - d = 5$; $d = 35$ cm

TEACHER CERTIFICATION STUDY GUIDE

24. The magnitude of a force is:
(Easy)

    A. Directly proportional to mass and inversely to acceleration

    B. Inversely proportional to mass and directly to acceleration

    C. Directly proportional to both mass and acceleration

    D. Inversely proportional to both mass and acceleration

Answer: C

To solve this problem, recall Newton's 2nd Law, i.e. net force is equal to mass times acceleration. Therefore, the only possible answer is (C).

25. A projectile with a mass of 1.0 kg has a muzzle velocity of 1500.0 m/s when it is fired from a cannon with a mass of 500.0 kg. If the cannon slides on a frictionless track, it will recoil with a velocity of ____ m/s.
(Rigorous)

    A. 2.4

    B. 3.0

    C. 3.5

    D. 1500

Answer: B

To solve this problem, apply Conservation of Momentum to the cannon-projectile system. The system is initially at rest, with total momentum of 0 kg m/s. Since the cannon slides on a frictionless track, we can assume that the net momentum stays the same for the system. Therefore, the momentum forward (of the projectile) must equal the momentum backward (of the cannon). Thus:

$p_{projectile} = p_{cannon}$
$m_{projectile} \, v_{projectile} = m_{cannon} \, v_{cannon}$
(1.0 kg)(1500.0 m/s) = (500.0 kg)(x)
x = 3.0 m/s
Only answer (B) matches these calculations.

26. A car (mass m_1) is driving at velocity v, when it smashes into an unmoving car (mass m_2), locking bumpers. Both cars move together at the same velocity. The common velocity will be given by: *(Rigorous)*

   A. m_1v/m_2

   B. m_2v/m_1

   C. $m_1v/(m_1 + m_2)$

   D. $(m_1 + m_2)v/m_1$

Answer: C

In this problem, there is an inelastic collision, so the best method is to assume that momentum is conserved. (Recall that momentum is equal to the product of mass and velocity.)
Therefore, apply Conservation of Momentum to the two-car system:
Momentum at Start = Momentum at End
(Mom. of Car 1) + (Mom. of Car 2) = (Mom. of 2 Cars Coupled)
$m_1v + 0 = (m_1 + m_2)x$
$x = m_1v/(m_1 + m_2)$
Only answer (C) matches these calculations.

Watch out for the other answers, because errors in algebra could lead to a match with incorrect answer (D), and assumption of an elastic collision could lead to a match with incorrect answer (A).

27. A satellite is in a circular orbit above the earth. Which statement is false?
   (Average Rigor)

   A. An external force causes the satellite to maintain orbit.

   B. The satellite's inertia causes it to maintain orbit.

   C. The satellite is accelerating toward the earth.

   D. The satellite's velocity and acceleration are not in the same direction.

Answer: B

To answer this question, recall that in circular motion, an object's inertia tends to keep it moving straight (tangent to the orbit), so a centripetal force (leading to centripetal acceleration) must be applied. In this case, the centripetal force is gravity due to the earth, which keeps the object in motion. Thus, (A), (C), and (D) are true, and (B) is the only false statement.

28. A 100 g mass revolving around a fixed point, on the end of a 0.5 meter string, circles once every 0.25 seconds. What is the magnitude of the centripetal acceleration?
    (Average Rigor)

    A. 1.23 m/s^2

    B. 31.6 m/s^2

    C. 100 m/s^2

    D. 316 m/s^2

Answer: D

The centripetal acceleration is equal to the product of the radius and the square of the angular frequency ω. In this case, ω is equal to 25.1 Hz. Squaring this value and multiplying by 0.5 m yields the result in answer D.

29. Which statement best describes the relationship of simple harmonic motion to a simple pendulum of length L, mass m and displacement of arc length s?
(Average Rigor)

    A. A simple pendulum cannot be modeled using simple harmonic motion

    B. A simple pendulum may be modeled using the same expression as Hooke's law for displacement s, but with a spring constant equal to the tension on the string

    C. A simple pendulum may be modeled using the same expression as Hooke's law for displacement s, but with a spring constant equal to m g/L

    D. A simple pendulum typically does not undergo simple harmonic motion

Answer: C

The force on a simple pendulum may be expressed approximately (when displacement s is small) according to the following equation:

$$F \approx -\frac{mg}{L} s$$

This expression has the same form as Hooke's law (F = -kx). Thus, answer C is the most correct response. Another approach to the question is to eliminate answers A and D as obviously incorrect, and then to eliminate answer B as not having appropriate units for the spring constant.

TEACHER CERTIFICATION STUDY GUIDE

30. A mass of 2 kg connected to a spring undergoes simple harmonic motion at a frequency of 3 Hz. What is the spring constant?
(Average Rigor)

    A. 6 kg/s^2

    B. 18 kg/s^2

    C. 710 kg/s^2

    D. 1000 kg/s^2

Answer: C

The spring constant, k, is equal to mω^2. In this case, ω is equal to 2π times the frequency of 3 Hz. The spring constant may be derived quickly by recognizing that the position of the mass varies sinusoidally with time at an angular frequency ω. Noting that the acceleration is the second derivative of the position with respect to time, the expression for k in Hooke's law (F = -kx) can be easily derived.

31. The kinetic energy of an object is _____ proportional to its _____.
(Average Rigor)

    A. Inversely...inertia

    B. Inversely...velocity

    C. Directly...mass

    D. Directly...time

Answer: C

To answer this question, recall that kinetic energy is equal to one-half of the product of an object's mass and the square of its velocity:
KE = ½ m v^2

Therefore, kinetic energy is directly proportional to mass, and the answer is (C). Note that although kinetic energy is associated with both velocity and momentum (a measure of inertia), it is not *inversely* proportional to either one.

TEACHER CERTIFICATION STUDY GUIDE

32. A force is given by the vector 5 N x + 3 N y (where x and y are the unit vectors for the x- and y- axes, respectively). This force is applied to move a 10 kg object 5 m, in the x direction. How much work was done?
(Rigorous)

    A. 250 J

    B. 400 J

    C. 40 J

    D. 25 J

Answer: D

To find out how much work was done, note that work counts only the force in the direction of motion. Therefore, the only part of the vector that we use is the 5 N in the x-direction. Note, too, that the mass of the object is not relevant in this problem. We use the work equation:
Work = (Force in direction of motion) (Distance moved)
Work = (5 N) (5 m)
Work = 25 J
This is consistent only with answer (D).

33. An office building entry ramp uses the principle of which simple machine?
(Easy)

    A. Lever

    B. Pulley

    C. Wedge

    D. Inclined Plane

Answer: D

To answer this question, recall the definitions of the various simple machines. A ramp, which trades a longer traversed distance for a shallower slope, is an example of an Inclined Plane, consistent with answer (D). Levers and Pulleys act to change size and/or direction of an input force, which is not relevant here. Wedges apply the same force over a smaller area, increasing pressure—again, not relevant in this case.

TEACHER CERTIFICATION STUDY GUIDE

34. If the internal energy of a system remains constant, how much work is done by the system if 1 kJ of heat energy is added?
    (Average Rigor)

    A. 0 kJ

    B. -1 kJ

    C. 1 kJ

    D. 3.14 kJ

Answer: C

According to the first law of thermodynamics, if the internal energy of a system remains constant, then any heat energy added to the system must be balanced by the system performing work on its surroundings. In the case of an ideal gas, the gas would necessarily expand when heated, assuming a constant internal energy was somehow maintained. Applying conservation of energy, answer C is found to be correct.

35. A calorie is the amount of heat energy that will:
    (Easy)

    A. Raise the temperature of one gram of water from 14.5° C to 15.5° C.

    B. Lower the temperature of one gram of water from 16.5° C to 15.5° C

    C. Raise the temperature of one gram of water from 32° F to 33° F

    D. Cause water to boil at two atmospheres of pressure.

Answer: A

The definition of a calorie is, "the amount of energy to raise one gram of water by one degree Celsius," and so answer (A) is correct. Do not get confused by the fact that 14.5° C seems like a random number. Also, note that answer (C) tries to confuse you with degrees Fahrenheit, which are irrelevant to this problem.

36. Use the information on heats below to solve this problem. An ice block at 0° Celsius is dropped into 100 g of liquid water at 18° Celsius. When thermal equilibrium is achieved, only liquid water at 0° Celsius is left. What was the mass, in grams, of the original block of ice?

   Given: Heat of fusion of ice = 80 cal/g
   Heat of vaporization of ice = 540 cal/g
   Specific Heat of ice = 0.50 cal/g°C
   Specific Heat of water = 1 cal/g°C

   (Rigorous)

   A. 2.0

   B. 5.0

   C. 10.0

   D. 22.5

Answer: D

To solve this problem, apply Conservation of Energy to the ice-water system. Any gain of heat to the melting ice must be balanced by loss of heat in the liquid water. Use the two equations relating temperature, mass, and energy:
$Q = m C \Delta T$ (for heat loss/gain from change in temperature)
$Q = m L$ (for heat loss/gain from phase change)
where Q is heat change; m is mass; C is specific heat; ΔT is change in temperature; L is heat of phase change (in this case, melting, also known as "fusion").

Then
$Q_{ice\ to\ water} = Q_{water\ to\ ice}$
(Note that the ice only melts; it stays at 0° Celsius—otherwise, we would have to include a term for warming the ice as well. Also the information on the heat of vaporization for water is irrelevant to this problem.)
$m L = m C \Delta T$
x (80 cal/g) = 100g 1cal/g°C 18°C
x (80 cal/g) = 1800 cal
x = 22.5 g

   Only answer (D) matches this result.

TEACHER CERTIFICATION STUDY GUIDE

37. Heat transfer by electromagnetic waves is termed:
(Easy)

    A. Conduction

    B. Convection

    C. Radiation

    D. Phase Change

Answer: C

To answer this question, recall the different ways that heat is transferred. Conduction is the transfer of heat through direct physical contact and molecules moving and hitting each other. Convection is the transfer of heat via density differences and flow of fluids. Radiation is the transfer of heat via electromagnetic waves (and can occur in a vacuum). Phase Change causes transfer of heat (though not of temperature) in order for the molecules to take their new phase. This is consistent, therefore, only with answer (C).

38. A cooking thermometer in an oven works because the metals it is composed of have different:
(Average Rigor)

    A. Melting points

    B. Heat convection

    C. Magnetic fields

    D. Coefficients of expansion

Answer: D

A thermometer of the type that can withstand oven temperatures works by having more than one metal strip. These strips expand at different rates with temperature increases, causing the dial to register the new temperature. This is consistent only with answer (D). If you did not know how an oven thermometer works, you could still omit the incorrect answers: It is unlikely that the metals in a thermometer would melt in the oven to display the temperature; the magnetic fields would not be useful information in this context; heat convection applies in fluids, not solids.

39. **Which of the following is not an assumption upon which the kinetic-molecular theory of gases is based?**
    (Rigorous)

    A. Quantum mechanical effects may be neglected

    B. The particles of a gas may be treated statistically

    C. The particles of the gas are treated as very small masses

    D. Collisions between gas particles and container walls are inelastic

Answer: D

Since the kinetic-molecular theory is classical in nature, quantum mechanical effects are indeed ignored, and answer A is incorrect. The theory also treats gases as a statistical collection of point-like particles with finite masses. As a result, answers B and C may also be eliminated. Thus, answer D is correct: collisions between gas particles and container walls are treated as elastic in the kinetic-molecular theory.

40. **What is temperature?**
    (Average Rigor)

    A. Temperature is a measure of the conductivity of the atoms or molecules in a material

    B. Temperature is a measure of the kinetic energy of the atoms or molecules in a material

    C. Temperature is a measure of the relativistic mass of the atoms or molecules in a material

    D. Temperature is a measure of the angular momentum of electrons in a material

Answer: B

Temperature is, in fact, a measure of the kinetic energy of the constituent components of a material. Thus, as a material is heated, the atoms or molecules that compose it acquire greater energy of motion. This increased motion results in the breaking of chemical bonds and in an increase in disorder, thus leading to melting or vaporizing of the material at sufficiently high temperatures.

41. Solids expand when heated because:
(Rigorous)

- A. Molecular motion causes expansion
- B. PV = nRT
- C. Magnetic forces stretch the chemical bonds
- D. All material is effectively fluid

Answer: A

When any material is heated, the heat energy becomes energy of motion for the material's molecules. This increased motion causes the material to expand (or sometimes to change phase). Therefore, the answer is (A). Answer (B) is the ideal gas law, which gives a relationship between temperature, pressure, and volume for gases. Answer (C) is a red herring (misleading answer that is untrue). Answer (D) may or may not be true, but it is not the best answer to this question.

42. What should be the behavior of an electroscope, which has been grounded in the presence of a positively charged object (1), after the ground connection is removed and then the charged object is removed from the vicinity (2)?
(Average Rigor)

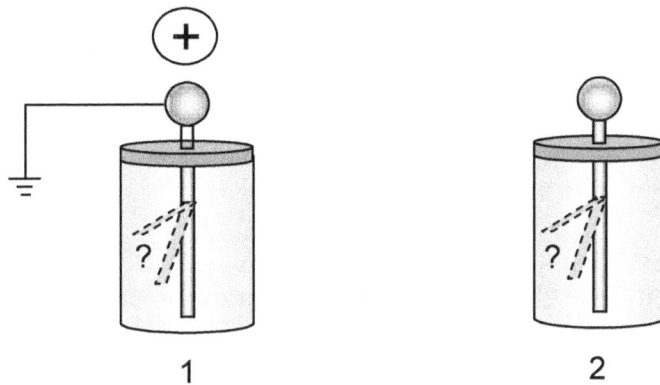

1
2

A. The metal leaf will start deflected (1) and then relax to an undeflected position (2)

B. The metal leaf will start in an undeflected position (1) and then be deflected (2)

C. The metal leaf will remain undeflected in both cases

D. The metal leaf will be deflected in both cases

Answer: B

When grounded, the electroscope will show no deflection. Nevertheless, if the ground is then removed and the charged object taken from the vicinity (in that order), the excess charge that existed near the sphere of the electroscope will distribute itself throughout the instrument, resulting in an overall net excess charge that will deflect the metal leaf.

43. The electric force in Newtons, on two small objects (each charged to –10 microCoulombs and separated by 2 meters) is:
    (Rigorous)

    A. 1.0

    B. 9.81

    C. 31.0

    D. 0.225

Answer: D

To answer this question, use Coulomb's Law, which gives the electric force between two charged particles:
$F = k Q_1 Q_2 / r^2$
Then our unknown is F, and our knowns are:
$k = 9.0 \times 10^9$ Nm^2/C^2
$Q_1 = Q_2 = -10 \times 10^{-6}$ C
$r = 2$ m

Therefore
$F = (9.0 \times 10^9)(-10 \times 10^{-6})(-10 \times 10^{-6})/(2^2)$ N
$F = 0.225$ N

This is compatible only with answer (D).

44. A 10 ohm resistor and a 50 ohm resistor are connected in parallel. If the current in the 10 ohm resistor is 5 amperes, the current (in amperes) running through the 50 ohm resistor is:
   (Rigorous)

   A. 1

   B. 50

   C. 25

   D. 60

Answer: A

To answer this question, use Ohm's Law, which relates voltage to current and resistance:
V = IR
where V is voltage; I is current; R is resistance.

We also use the fact that in a parallel circuit, the voltage is the same across the branches.

Because we are given that in one branch, the current is 5 amperes and the resistance is 10 ohms, we deduce that the voltage in this circuit is their product, 50 volts (from V = IR).

We then use V = IR again, this time to find I in the second branch. Because V is 50 volts, and R is 50 ohm, we calculate that I has to be 1 ampere.

This is consistent only with answer (A).

45. How much power is dissipated through the following resistive circuit?
(Average Rigor)

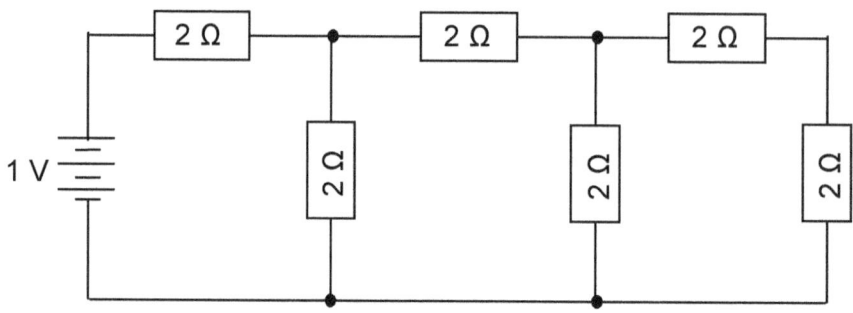

    A. 0 W

    B. 0.22 W

    C. 0.31 W

    D. 0.49 W

Answer: C

Use the rules of series and parallel resistors to quickly form an equivalent circuit with a single voltage source and a single resistor. In this case, the equivalent resistance is 3.25 Ω. The power dissipated by the circuit is the square of the voltage divided by the resistance. The final answer is C.

46. **The greatest number of 100 watt lamps that can be connected in parallel with a 120 volt system without blowing a 5 amp fuse is:**
    (Rigorous)

    A. 24

    B. 12

    C. 6

    D. 1

Answer: C

To solve fuse problems, you must add together all the drawn current in the parallel branches, and make sure that it is less than the fuse's amp measure. Because we know that electrical power is equal to the product of current and voltage, we can deduce that:
I = P/V (I = current (amperes); P = power (watts); V = voltage (volts))

Therefore, for each lamp, the current is 100/120 amperes, or 5/6 ampere. The highest possible number of lamps is thus six, because six lamps at 5/6 ampere each adds to 5 amperes; more will blow the fuse.

This is consistent only with answer (C).

47. **Which of the following statements may be taken as a legitimate inference based upon the Maxwell equation that states $\nabla \cdot \mathbf{B} = 0$?**
    (Average Rigor)

    A. The electric and magnetic fields are decoupled

    B. The electric and magnetic fields are mediated by the W boson

    C. There are no photons

    D. There are no magnetic monopoles

Answer: D

Since the divergence of the magnetic flux density is always zero, there cannot be any magnetic monopoles (charges), given this Maxwell equation. If Gauss's law is applied to magnetic flux in the same manner as it is to electric flux, then the total magnetic "charge" contained within any closed surface must always be zero. This is another way of viewing the problem. Thus, answer D is correct. This answer may also be chosen by elimination of the other statements, which are untenable.

48. What effect might an applied external magnetic field have on the magnetic domains of a ferromagnetic material?
(Rigorous)

   A. The domains that are not aligned with the external field increase in size, but those that are aligned decrease in size

   B. The domains that are not aligned with the external field decrease in size, but those that are aligned increase in size

   C. The domains align perpendicular to the external field

   D. There is no effect on the magnetic domains

Answer: B

Recall that ferromagnetic domains are portions of a magnetic material that have a local magnetic moment. The material may have an overall lack of a magnetic moment due to random alignment of its domains. In the presence of an applied field, the domains may align with the field to some extent, or the boundaries of the domains may shift to give greater weight to those domains that are aligned with the field, at the expense of those domains that are not aligned with the field. As a result, of the possibilities above, B is the best answer.

49. What is the effect of running current in the same direction along two parallel wires, as shown below?
(Rigorous)

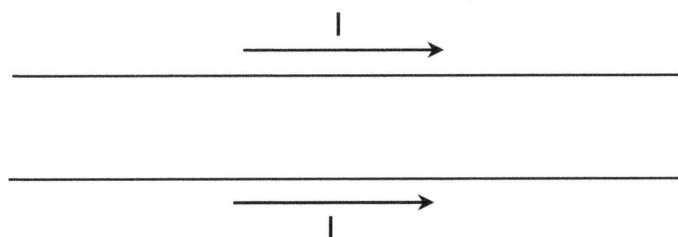

   A. There is no effect

   B. The wires attract one another

   C. The wires repel one another

   D. A torque is applied to both wires

Answer: B

Since the direction of the force on a current element is proportional to the cross product of the direction of the current element and the magnetic field, there is either an attractive or repulsive force between the two wires shown above. Using the right hand rule, it can be found that the magnetic field on the top wire due to the bottom wire is directed out of the plane of the page. Performing the cross product shows that the force on the upper wire is directed toward the lower wire. A similar argument can be used for the lower wire. Thus, the correct answer is B: an attractive force is exerted on the wires.

TEACHER CERTIFICATION STUDY GUIDE

50. The current induced in a coil is defined by which of the following laws?
    (Easy)

    A. Lenz's Law

    B. Burke's Law

    C. The Law of Spontaneous Combustion

    D. Snell's Law

Answer: A

Lenz's Law states that an induced electromagnetic force always gives rise to a current whose magnetic field opposes the original flux change. There is no relevant "Snell's Law," "Burke's Law," or "Law of Spontaneous Combustion" in electromagnetism. (In fact, only Snell's Law is a real law of these three, and it refers to refracted light.) Therefore, the only appropriate answer is (A).

51. A light bulb is connected in series with a rotating coil within a magnetic field. The brightness of the light may be increased by any of the following except:
    (Average Rigor)

    A. Rotating the coil more rapidly.

    B. Using more loops in the coil.

    C. Using a different color wire for the coil.

    D. Using a stronger magnetic field.

Answer: C

To answer this question, recall that the rotating coil in a magnetic field generates electric current, by Faraday's Law. Faraday's Law states that the amount of emf generated is proportional to the rate of change of magnetic flux through the loop. This increases if the coil is rotated more rapidly (A), if there are more loops (B), or if the magnetic field is stronger (D). Thus, the only answer to this question is (C).

52. **What is the direction of the magnetic field at the center of the loop of current (I) shown below (i.e., at point A)?**
(Easy)

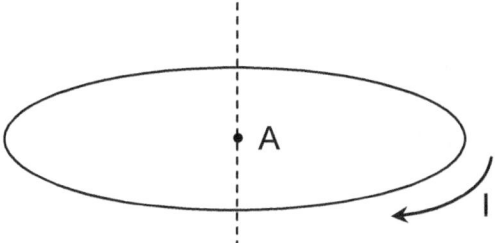

A. Down, along the axis (dotted line)

B. Up, along the axis (dotted line)

C. The magnetic field is oriented in a radial direction

D. There is no magnetic field at point A

Answer: A

The magnetic field may be found by applying the right-hand rule. The magnetic field curls around the wire in the direction of the curled fingers when the thumb is pointed in the direction of the current. Since there is a degree of symmetry, with point A lying in the center of the loop, the contributions of all the current elements on the loop must yield a field that is either directed up or down at the axis. Use of the right-hand rule indicates that the field is directed down. Thus, answer A is correct.

TEACHER CERTIFICATION STUDY GUIDE

53. **The use of two circuits next to each other, with a change in current in the primary circuit, demonstrates:**
    (Rigorous)

    A. Mutual current induction

    B. Dielectric constancy

    C. Harmonic resonance

    D. Resistance variation

Answer: A

To answer this question, recall that changing current induces a change in magnetic flux, which in turn causes a change in current to oppose that change (Lenz's and Faraday's Laws). Thus, (A) is correct. If you did not remember that, note that harmonic resonance is irrelevant here (eliminating (C)), and there is no change in resistance in the circuits (eliminating (D)).

54. **A semi-conductor allows current to flow:**
    (Easy)

    A. Never

    B. Always

    C. As long as it stays below a maximum temperature

    D. When a minimum voltage is applied

Answer: D

To answer this question, recall that semiconductors do not conduct as well as conductors (eliminating answer (B)), but they conduct better than insulators (eliminating answer (A)). Semiconductors can conduct better when the temperature is higher (eliminating answer (C)), and their electrons move most readily under a potential difference. Thus the answer can only be (D).

PHYSICS

55. **All of the following use semi-conductor technology, except a(n):**
    (Average Rigor)

    A. Transistor

    B. Diode

    C. Capacitor

    D. Operational Amplifier

Answer: C

Semi-conductor technology is used in transistors and operational amplifiers, and diodes are the basic unit of semi-conductors. Therefore the only possible answer is (C), and indeed a capacitor does not require semi-conductor technology.

56. **A wave generator is used to create a succession of waves. The rate of wave generation is one every 0.33 seconds. The period of these waves is:**
    (Average Rigor)

    A. 2.0 seconds

    B. 1.0 seconds

    C. 0.33 seconds

    D. 3.0 seconds

Answer: C

The definition of a period is the length of time between wave crests. Therefore, when waves are generated one per 0.33 seconds, that same time (0.33 seconds) is the period. This is consistent only with answer (C). Do not be trapped into calculating the number of waves per second, which might lead you to choose answer (D).

57. **An electromagnetic wave propagates through a vacuum. Independent of its wavelength, it will move with constant:**
    (Easy)

    A. Acceleration

    B. Velocity

    C. Induction

    D. Sound

Answer: B

Electromagnetic waves are considered always to travel at the speed of light, so answer (B) is correct. Answers (C) and (D) can be eliminated in any case, because induction is not relevant here, and sound does not travel in a vacuum.

58. **A wave has speed 60 m/s and wavelength 30,000 m. What is the frequency of the wave?**
    (Average Rigor)

    A. 2.0×10^{-3} Hz

    B. 60 Hz

    C. 5.0×10^{2} Hz

    D. 1.8×10^{6} Hz

Answer: A

To answer this question, recall that wave speed is equal to the product of wavelength and frequency. Thus:
60 m/s = (30,000 m) (frequency)
frequency = 2.0×10^{-3} Hz

This is consistent only with answer (A).

59. Rainbows are created by:
(Easy)

- A. Reflection, dispersion, and recombination
- B. Reflection, resistance, and expansion
- C. Reflection, compression, and specific heat
- D. Reflection, refraction, and dispersion

Answer: D

To answer this question, recall that rainbows are formed by light that goes through water droplets and is dispersed into its colors. This is consistent with both answers (A) and (D). Then note that refraction is important in bending the differently colored light waves, while recombination is not a relevant concept here. Therefore, the answer is (D).

60. Which of the following is *not* a legitimate explanation for refraction of light rays at boundaries between different media?
(Rigorous)

- A. Light seeks the path of least time between two different points
- B. Due to phase matching and other boundary conditions, plane waves travel in different directions on either side of the boundary, depending on the material parameters
- C. The electric and magnetic fields become decoupled at the boundary
- D. Light rays obey Snell's law

Answer: C

Even if the exact implications of each explanation are not known or understood, answer C can be chosen due to its plain incorrectness. The other responses involve more or less fundamental explanations for the refraction of light rays (which are equivalent to plane waves) at media boundaries.

61. A stationary sound source produces a wave of frequency *F*. An observer at position A is moving toward the horn, while an observer at position B is moving away from the horn. Which of the following is true?
(Rigorous)

    A. $F_A < F < F_B$

    B. $F_B < F < F_A$

    C. $F < F_A < F_B$

    D. $F_B < F_A < F$

Answer: B

To answer this question, recall the Doppler Effect. As a moving observer approaches a sound source, s/he intercepts wave fronts sooner than if s/he were standing still. Therefore, the wave fronts seem to be coming more frequently. Similarly, as an observer moves away from a sound source, the wave fronts take longer to reach him/her. Therefore, the wave fronts seem to be coming less frequently. Because of this effect, the frequency at B will seem lower than the original frequency, and the frequency at A will seem higher than the original frequency. The only answer consistent with this is (B). Note also, that even if you weren't sure of which frequency should be greater/smaller, you could still reason that A and B should have opposite effects, and be able to eliminate answer choices (C) and (D).

62. A monochromatic ray of light passes from air to a thick slab of glass (n = 1.41) at an angle of 45° from the normal. At what angle does it leave the air/glass interface?
    (Rigorous)

    A. 45°

    B. 30°

    C. 15°

    D. 55°

Answer: B

To solve this problem use Snell's Law:
$n_1 \sin\theta_1 = n_2 \sin\theta_2$ (where n_1 and n_2 are the indexes of refraction and θ_1 and θ_2 are the angles of incidence and refraction).

Then, since the index of refraction for air is 1.0, we deduce:
$1 \sin 45° = 1.41 \sin x$
$x = \sin^{-1}((1/1.41) \sin 45°)$
$x = 30°$

This is consistent only with answer (B). Also, note that you could eliminate answers (A) and (D) in any case, because the refracted light will have to bend at a smaller angle when entering glass.

63. If one sound is ten decibels louder than another, the ratio of the intensity of the first to the second is:
(Average Rigor)

    A. 20:1

    B. 10:1

    C. 1:1

    D. 1:10

Answer: B

To answer this question, recall that a decibel is defined as ten times the log of the ratio of sound intensities:
(decibel measure) = 10 log (I / I_0) where I_0 is a reference intensity.

Therefore, in our case,
(decibels of first sound) = (decibels of second sound) + 10
10 log (I_1 / I_0) = 10 log (I_2 / I_0) + 10
10 log I_1 − 10 log I_0 = 10 log I_2 − 10 log I_0 + 10
10 log I_1 − 10 log I_2 = 10
log (I_1 / I_2) = 1
I_1 / I_2 = 10

This is consistent only with answer (B).
(Be careful not to get the two intensities confused with each other.)

TEACHER CERTIFICATION STUDY GUIDE

64. The velocity of sound is greatest in:
(Average Rigor)

    A. Water

    B. Steel

    C. Alcohol

    D. Air

Answer: B

Sound is a longitudinal wave, which means that it shakes its medium in a way that propagates as sound traveling. The speed of sound depends on both elastic modulus and density, but for a comparison of the above choices, the answer is always that sound travels faster through a solid like steel, than through liquids or gases. Thus, the answer is (B).

65. A vibrating string's frequency is _____ proportional to the _____.
(Rigorous)

    A. Directly; Square root of the tension

    B. Inversely; Length of the string

    C. Inversely; Squared length of the string

    D. Inversely; Force of the plectrum

Answer: A

To answer this question, recall that
$f = (n v) / (2 L)$ where f is frequency; v is velocity; L is length

and

$v = (F_{tension} / (m / L))^{1/2}$ where $F_{tension}$ is tension; m is mass; others as above

so

$f = (n / 2 L) ((F_{tension} / (m / L))^{1/2})$

indicating that frequency is directly proportional to the square root of the tension force. This is consistent only with answer (A). Note that in the final frequency equation, there is an inverse relationship with the square root of the length (after canceling like terms). This is not one of the options, however.

TEACHER CERTIFICATION STUDY GUIDE

66. Which of the following apparatus can be used to measure the wavelength of a sound produced by a tuning fork?
(Average Rigor)

      A. A glass cylinder, some water, and iron filings

      B. A glass cylinder, a meter stick, and some water

      C. A metronome and some ice water

      D. A comb and some tissue

Answer: B

To answer this question, recall that a sound will be amplified if it is reflected back to cause positive interference. This is the principle behind musical instruments that use vibrating columns of air to amplify sound (e.g. a pipe organ). Therefore, presumably a person could put varying amounts of water in the cylinder, and hold the vibrating tuning fork above the cylinder in each case. If the tuning fork sound is amplified when put at the top of the column, then the length of the air space would be an integral multiple of the sound's wavelength. This experiment is consistent with answer (B). Although the experiment would be tedious, none of the other options for materials suggest a better alternative.

67. The highest energy is associated with:
(Easy)

      A. UV radiation

      B. Yellow light

      C. Infrared radiation

      D. Gamma radiation

Answer: D

To answer this question, recall the electromagnetic spectrum. The highest energy (and therefore frequency) rays are those with the lowest wavelength, i.e. gamma rays. (In order of frequency from lowest to highest are: radio, microwave, infrared, red through violet visible light, ultraviolet, X-rays, gamma rays.) Thus, the only possible answer is (D). Note that even if you did not remember the spectrum, you could deduce that gamma radiation is considered dangerous and thus might have the highest energy.

TEACHER CERTIFICATION STUDY GUIDE

68. An object two meters tall is speeding toward a plane mirror at 10 m/s. What happens to the image as it nears the surface of the mirror?
(Rigorous)

    A. It becomes inverted.

    B. The Doppler Effect must be considered.

    C. It remains two meters tall.

    D. It changes from a real image to a virtual image.

Answer: C

Note that the mirror is a plane mirror, so the image is always a virtual image of the same size as the object. If the mirror were concave, then the image would be inverted until the object came within the focal distance of the mirror. The Doppler Effect is not relevant here. Thus, the only possible answer is (C).

69. Automobile mirrors that have a sign, "objects are closer than they appear" say so because:
(Rigorous)

    A. The real image of an obstacle, through a converging lens, appears farther away than the object.

    B. The real or virtual image of an obstacle, through a converging mirror, appears farther away than the object.

    C. The real image of an obstacle, through a diverging lens, appears farther away than the object.

    D. The virtual image of an obstacle, through a diverging mirror, appears farther away than the object.

Answer: D

To answer this question, first eliminate answer choices (A) and (C), because we have a mirror, not a lens. Then draw ray diagrams for diverging (convex) and converging (concave) mirrors, and note that because the focal point of a diverging mirror is behind the surface, the image is smaller than the object. This creates the illusion that the object is farther away, and therefore (D) is the correct answer.

70. If an object is 20 cm from a convex lens whose focal length is 10 cm, the image is:
    (Rigorous)

    A. Virtual and upright

    B. Real and inverted

    C. Larger than the object

    D. Smaller than the object

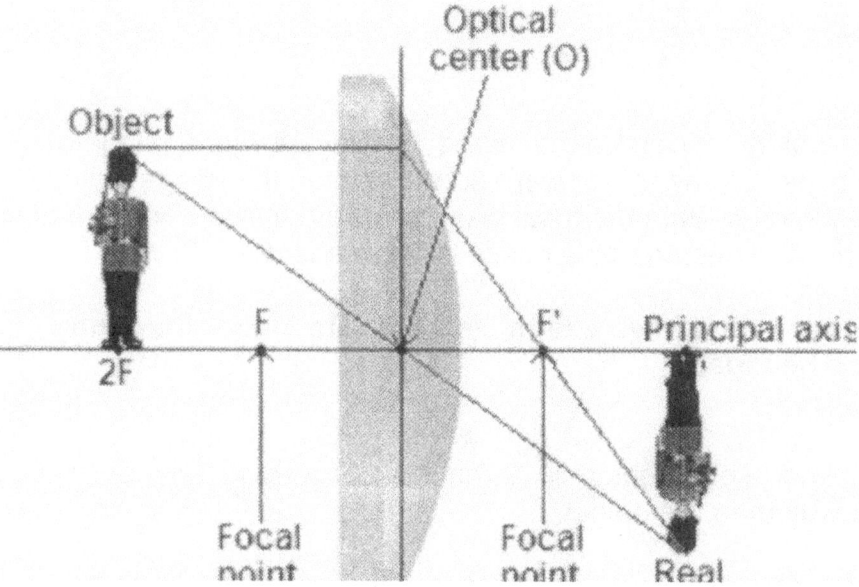

Answer: B

To solve this problem, draw a lens diagram with the lens, focal length, and image size.

The ray from the top of the object straight to the lens is focused through the far focus point; the ray from the top of the object through the near focus goes straight through the lens; the ray from the top of the object through the center of the lens continues. These three meet to form the "top" of the image, which is therefore real and inverted. This is consistent only with answer (B).

71. **The constant of proportionality between the energy and the frequency of electromagnetic radiation is known as the:**
    (Easy)

    A. Rydberg constant

    B. Energy constant

    C. Planck constant

    D. Einstein constant

Answer: C

Planck estimated his constant to determine the ratio between energy and frequency of radiation. The Rydberg constant is used to find the wavelengths of the visible lines on the hydrogen spectrum.

The other options are not relevant options, and may not actually have physical meaning. Therefore, the only possible answer is (C).

72. **Which phenomenon was first explained using the concept of quantization of energy, thus providing one of the key foundational principles for the later development of quantum theory?**
    (Rigorous)

    A. The photoelectric effect

    B. Time dilation

    C. Blackbody radiation

    D. Magnetism

Answer: C

Although the photoelectric effect applied principles of quantization in explaining the behavior of electrons emitted from a metallic surface when the surface is illuminated with electromagnetic radiation, the explanation of the phenomenon of blackbody radiation, provided by Max Planck, was the first major success of the concept of quantized energy. Magnetism may be explained quantum mechanically, but such an explanation was not forthcoming until well after Planck's quantization hypothesis. Time dilation is primarily explained through relativity theory.

TEACHER CERTIFICATION STUDY GUIDE

73. Which statement best describes why population inversion is necessary for a laser to operate?
(Rigorous)

    A. Population inversion prevents too many electrons from being excited into higher energy levels, thus preventing damage to the gain medium.

    B. Population inversion maintains a sufficient number of electrons in a higher energy state so as to allow a significant amount of stimulated emission.

    C. Population inversion prevents the laser from producing coherent light.

    D. Population inversion is not necessary for the operation of most lasers.

Answer: B

Population inversion is a state in which there are a larger number of electrons in a particular higher-energy excited state than in a particular lower-energy state. When perturbed by a passing photon, these electrons may then emit a photon of the same energy (frequency) and phase. This is the process of stimulated emission, which, when population inversion is obtained, can produce something of a "chain reaction," thus giving lasers their characteristically monochromatic and highly coherent light.

74. Bohr's theory of the atom was the first to quantize:
(Average Rigor)

    A. Work

    B. Angular Momentum

    C. Torque

    D. Duality

Answer: B

Bohr was the first to quantize the angular momentum of electrons, as he combined Rutherford's planet-style model with his knowledge of emerging quantum theory. Recall that he derived a "quantum condition" for the single electron, requiring electrons to exist at specific energy levels

75. Two neutral isotopes of a chemical element have the same numbers of:
(Easy)

- A. Electrons and Neutrons
- B. Electrons and Protons
- C. Protons and Neutrons
- D. Electrons, Neutrons, and Protons

Answer: B

To answer this question, recall that isotopes vary in their number of neutrons. (This fact alone eliminates answers (A), (C), and (D).) If you did not recall that fact, note that we are given that the two samples are of the same element, constraining the number of protons to be the same in each case. Then, use the fact that the samples are neutral, so the number of electrons must exactly balance the number of protons in each case. The only correct answer is thus (B).

76. When a radioactive material emits an alpha particle only, its atomic number will:
(Average Rigor)

- A. Decrease
- B. Increase
- C. Remain unchanged
- D. Change randomly

Answer: A

To answer this question, recall that in alpha decay, a nucleus emits the equivalent of a Helium atom. This includes two protons, so the original material changes its atomic number by a decrease of two.

TEACHER CERTIFICATION STUDY GUIDE

77. **Ten grams of a sample of a radioactive material (half-life = 12 days) were stored for 48 days and re-weighed. The new mass of material was:**
    (Rigorous)

    A. 1.25 g

    B. 2.5 g

    C. 0.83 g

    D. 0.625 g

Answer: D

To answer this question, note that 48 days is four half-lives for the material. Thus, the sample will degrade by half four times. At first, there are ten grams, then (after the first half-life) 5 g, then 2.5 g, then 1.25 g, and after the fourth half-life, there remains 0.625 g. You could also do the problem mathematically, by multiplying ten times $(½)^4$, i.e. ½ for each half-life elapsed.

78. **Which of the following pairs of elements are not found to fuse in the centers of stars?**
    (Average Rigor)

    A. Oxygen and Helium

    B. Carbon and Hydrogen

    C. Beryllium and Helium

    D. Cobalt and Hydrogen

Answer: D

To answer this question, recall that fusion is possible only when the final product has more binding energy than the reactants. Because binding energy peaks near a mass number of around 56, corresponding to Iron, any heavier elements would be unlikely to fuse in a typical star. (In very massive stars, there may be enough energy to fuse heavier elements.) Of all the listed elements, only Cobalt is heavier than iron, so answer (D) is correct.

79. In a fission reactor, heavy water:
(Average Rigor)

    A. Cools off neutrons to control temperature

    B. Moderates fission reactions

    C. Initiates the reaction chain

    D. Dissolves control rods

Answer: B

In a nuclear reactor, heavy water is made up of oxygen atoms with hydrogen atoms called 'deuterium,' which contain two neutrons each. This allows the water to slow down (moderate) the neutrons, without absorbing many of them. This is consistent only with answer (B).

80. Given the following values for the masses of a proton, a neutron and an alpha particle, what is the nuclear binding energy of an alpha particle?
(Rigorous)

Proton mass = 1.6726×10^{-27} kg
Neutron mass = 1.6749×10^{-27} kg
Alpha particle mass = 6.6465×10^{-27} kg

    A. 0 J

    B. 7.3417×10^{-27} J

    C. 4 J

    D. 4.3589×10^{-12} J

Answer: D

The nuclear binding energy is the amount of energy that is required to break the nucleus into its component nucleons. In this case, the binding energy of an alpha particle, which is composed of two protons and two neutrons, is calculated by first finding the difference between the sum of the masses of all the nucleons and the mass of the alpha particle. Using the equation $E = mc^2$ to find the energy in terms of the mass difference of 4.85×10^{-29} kg, and using the speed of light of about 2.9979×10^8 m/s, the result is the value given in answer D.

www.ingramcontent.com/pod-product-compliance
Lightning Source LLC
Chambersburg PA
CBHW080535300426
44111CB00017B/2736